REBUILD & POWERTUNE CARTER/EDELBROCK CARBURETORS

Covers AFB, AVS and TQ Models for Street, Performance and Racing

Larry Shepard

HPBOOKS

HPBooks

Published by the Penguin Group
Penguin Group (USA) Inc.
375 Hudson Street, New York, New York 10014, USA
Penguin Group (Canada), 90 Eglinton Avenue East, Suite 700, Toronto, Ontario M4P 2Y3, Canada
(a division of Pearson Penguin Canada Inc.)
Penguin Books Ltd., 80 Strand, London WC2R 0RL, England
Penguin Group Ireland, 25 St. Stephen's Green, Dublin 2, Ireland (a division of Penguin Books Ltd.)
Penguin Group (Australia), 250 Camberwell Road, Camberwell, Victoria 3124, Australia
(a division of Pearson Australia Group Pty. Ltd.)
Penguin Books India Pvt. Ltd., 11 Community Centre, Panchsheel Park, New Delhi—110 017, India
Penguin Group (NZ), 67 Apollo Drive, Rosedale, North Shore 0632, New Zealand
(a division of Pearson New Zealand Ltd.)
Penguin Books (South Africa) (Pty.) Ltd., 24 Sturdee Avenue, Rosebank, Johannesburg 2196, South Africa

Penguin Books Ltd., Registered Offices: 80 Strand, London WC2R 0RL, England

While the author has made every effort to provide accurate telephone numbers and Internet addresses at the time of publication, neither the publisher nor the author assumes any responsibility for errors, or for changes that occur after publication. Further, the publisher does not have any control over and does not assume any responsibility for author or third-party websites or their content.

REBUILD & POWERTUNE CARTER/EDELBROCK CARBURETORS

Copyright © 2010 by Larry Shepard
Cover design by Bird Studios
Front cover photo courtesy Edelbrock Corporation
Interior photos by author unless otherwise noted

All rights reserved. No part of this book may be reproduced, scanned, or distributed in any printed or electronic form without permission. Please do not participate in or encourage piracy of copyrighted materials in violation of the author's rights. Purchase only authorized editions.
HPBooks is a trademark of Penguin Group (USA) Inc.

First edition: January 2010

ISBN: 978-1-55788-555-5

PRINTED IN THE UNITED STATES OF AMERICA

10 9 8 7 6 5 4

NOTICE: The information in this book is true and complete to the best of our knowledge. All recommendations on parts and procedures are made without any guarantees on the part of the author or the publisher. Tampering with, altering, modifying, or removing any emissions-control device is a violation of federal law. Author and publisher disclaim all liability incurred in connection with the use of this information. We recognize that some words, engine names, model names, and designations mentioned in this book are the property of the trademark holder and are used for identification purposes only. This is not an official publication.

CONTENTS

Acknowledgments — iv
Introduction — v

Chapter 1
Carter History, Basic Theory & Identification — 1

Chapter 2
Carter/Edelbrock Components & Tech Specs — 17

Chapter 3
Carb Selection — 31

Chapter 4
Carb Rebuilding — 41

Chapter 5
Carb Hardware & Installation — 98

Chapter 6
Carb Adjustments & Related Hardware — 117

Chapter 7
Carb Troubleshooting, Tune-Up & Calibrations — 139

Chapter 8
Carb Emissions, Fuel Economy & Power — 151

Chapter 9
Racing & Special Applications — 163

ACKNOWLEDGMENTS

Writing a book of this size is certainly not a one-man job. Over the years I've had a lot of help. There are hundreds, perhaps thousands, of racers who have been involved in various race programs over the past forty years who have had some hand in developing carburetor technology. This book has its roots in the Chrysler Drag Racing seminars and the various tech support lines that have all helped customers solve problems. I've spent many hours at race tracks, SEMA and PRI trade shows participating in the displays done by Jack McCormack of McCormack Motorsports and the Mopar Road Show by Kirk Carbary. I've also made a lot of trips to the Chrysler Historical Museum.

I learned most of what I know about carburetors from John Bauman at the beginning of my career, then from Al Nichols during the last thirty-five years. This book is a reflection of their combined knowledge and experience. I greatly appreciate their time and patience.

Arrow Racing and Ray Barton Racing helped me with staging many of the photos in this book. In addition, many aftermarket manufacturers and suppliers, including Edelbrock, contributed photos as well.

A very special thanks to Bill Hancock for all the effort and assistance that he has given me on this project.

And perhaps most of all, we all owe a great deal of thanks to my editor, Michael Lutfy, who made all of this technical information readable.

Finally, a special thanks to my wife, Linda, for her enduring support and patience with my ever-mounting piles of research material and photographs.

INTRODUCTION

In this book, I will discuss the carburetor. The dictionary defines a carburetor as a device for mixing fuel with air to produce an explosive mixture for an internal combustion engine, once per cycle. Up until about the early 1980s most internal combustion engines were equipped with a carburetor. Most carbs were either side-draft, updraft or downdraft style. The updraft carburetor was only used on a few models from the earliest era. The side-draft carburetor was used frequently in Europe but was not popular on U.S. production models. By far, the most popular carburetor in the U.S. has been the downdraft style. This style tends to mean that the carb sits on top of the engine. The next major division in carburetors is the number of throttle bores. There are one-barrel, two-barrel and four-barrel versions and multiple carburetor versions that have six-barrel and eight-barrel configurations. The most popular production configurations were the two-barrel carb and the four-barrel carb. And the most popular race carburetor is the four-barrel. There are some three-barrel versions where the two secondary barrels are joined into one barrel making three that are designed for racing.

In the first fifty years of American production automobiles and engines, there were several carburetor manufacturers, but as engine performance became a bigger issue by the mid-1950s, the manufacturing field narrowed down to about four major players—Carter, Holley, Rochester (GM) and Ford/Autolite. While many of the basics are similar among these various carburetor manufacturers, and the technical theories and terms may overlap, details and specific photos are unique. Therefore I will concentrate on the Carter carburetors and even more specifically the models designated AFB (aluminum four-barrel), AVS (air valve secondary) and TQ (Thermo-Quad) in this book, all made by Carter.

The engineers that developed the early production engines had three main parameters to focus on—performance, drivability and fuel economy. Drivability was perhaps the most important feature prior to 1955. Performance, which is the engine's horsepower and torque output, became a major concern in 1955, but mainly on the performance models. Fuel economy was a concern on smaller engines and was popularized by the Mobil Gas Economy Run and the Union Pure Trials that were run in the 1960s. The original three-pronged development plan was widened to four prongs with the addition of emissions in 1967. There are three major steps in the

The typical Carter/Edelbrock four-barrel carburetor shown in a front view. It's an AVS made by Edelbrock.

This is the Chrysler 354–392 Firepower Hemi with two four-barrel carburetors inline, produced from about 1956 through 1958. The carbs are WCFB.

Edelbrock's version of the AVS, shown from the left side or driver's side. Note the carburetor linkage and the accelerator pump located just ahead (to left) of the carb linkage center pivot, located at bottom center.

emissions regulations: the 1967 emissions, the 1972 emissions and the OBD I and II (onboard diagnostics level 1 and 2). Each one of these plateaus was created by federal laws. Perhaps the most important one occurred in 1972 because it included unleaded gas requirements.

In a very general sense, the carburetor is a simple device but it is very important to the engine and a key to the engine's operation in almost every phase. By general description, the carburetor has a somewhat simple job to do—mix fuel with air. How the carburetor does this mixing job is not obvious once the carb is assembled. One of my main goals in the following chapters of this book is to remove the mystery that often surrounds the carburetor and try to decrease the common fears relating to the disassembly of the carburetor itself. Typically, these Carter four-barrel carburetors are held together by 8–10 screws. Once you understand the terms and features, the carburetor is easy to work on and simple to adjust. Early on in Chapter 1, I will discuss the term air/fuel ratio and what it means. The related rich/lean aspects of engine performance and how to achieve them, which are the keys to the success of your project, will be the indicators for the calibration process described in Chapter 6 and Chapter 7. I will keep referring to A/F ratio once the carburetor is installed onto the engine.

By the mid-1990s, multi-point fuel injection had replaced the carburetor in production applications. This change was mainly due to federal emissions standards that became standard with OBD I and II that required an onboard computer. Today we are used to fuel injection, but that wasn't true in the late '70s and early '80s. The induction story is usually told going from the carburetor to fuel injection because that was the way it happened chronologically. There were several steps in between, including electronically controlled carburetors and throttle-body injection that were utilized before we arrived at today's MPI—multi-point injection. So if we look at this transition in reverse, going from today back to yesterday, the job of the engine's computer, injectors, throttle body, perhaps twenty sensors and two TDC pickups is done by one carburetor! The carburetor is simple, easy to adjust and understand and is low in cost by comparison.

The key to understanding the carburetor's overall operation is to break down the engine's in-vehicle operation into three phases—idle, part throttle and wide-open throttle (WOT). Then you apply the air/fuel ratio

This is a 1972 and newer Thermo-Quad four-barrel carburetor. Note the two vertical pipes in the center of the carb and somewhat to the left in the photo. The '71 version has only one pipe and it is much larger in diameter.

The top view of the AVS carburetor has the primaries toward the bottom in this photo. The choke plate is vertical (off) and you can see directly into the primaries. The air valve is closed and shuts off the top of carburetor so you can't see the secondaries.

requirements to each phase (rich-or-lean) and adjust for improved performance. Learning how to adjust your own carburetor on your specific engine, the DIY approach—do-it-yourself—will save you a tremendous amount of money versus sending it out to a pro. It will also save you a lot of time because you can do it right now. However, a word of caution: When adjusting your carb, you should keep good records and write everything down—it is very easy to get lost and may not be easy to remember which setup was best.

—*Larry Shepard*

Chapter 1
Carter History, Basic Theory & Identification

The front view of a production AFB four-barrel carburetor. Note that the throttle linkage and accelerator pump are located on the right side of the photo (driver's side as installed).

The carburetor is a device designed to mix fuel with air to produce an explosive mixture for internal combustion engines. While there are side-draft and updraft carburetors, I will focus on the more common downdraft carburetors.

As the name implies, a downdraft carburetor is typically located on top of the engine. Although you could pour fuel directly into the intake manifold, the engine wouldn't run very well because the fuel isn't mixed or regulated. For proper engine operation, the air/fuel mixture must be mixed and regulated, and that is where the carb comes into play.

There are three main parameters to consider for production engines—performance, drivability and fuel economy. There are systems in the carburetor that focus on each specific aspect. The fuel mixing and the fuel regulation functions are designed in, so I will only cover the highlights. Our focus in this chapter is to offer some background on the Carter family of carburetors, and to become familiar with the terms we'll be using when discussing carbs in later chapters.

Carburetors can have more than one throttle bore or barrel. There are carbs with one barrel, two barrels and four barrels. The one-barrel will be used to explain the basic functions and operation of a carburetor. However, this book will focus on four-barrel carbs. The typical four-barrel carburetor has two primary and two secondary barrels or throttle bores. The typical primary bore size is either equal to the secondary or smaller. The left and right sides of the carburetor are generally identical.

There are many terms that are used in carburetor discussions. Some terms, like throttle bores and venturi, refer

The front view of the production AVS four-barrel carburetor. Note the single idle adjustment screw in the center of the carb, halfway between top and bottom. It has a left-hand thread.

directly to the carburetor itself while others, like cubic feet per minute (cfm), pressure, brake specific fuel consumption (BSFC) and air/fuel (A/F) ratio, refer to measured aspects of the carburetor's performance. I will discuss these terms in more detail shortly.

The front view of the 1972 and newer Thermo-Quad carburetor. Note that there are a lot more vacuum attachment nipples in the base-plate than the AFB and AVS versions.

The first racing eight-barrel tunnel ram system was used on the NHRA C/A 354 Hemi in the late 1950s and used two WCFB four-barrel carbs.

This is the mid to late 1950s Chrysler Firepower 354–392 Hemi V8 with dual four-barrel WCFB carbs inline. The 354–392 is now considered Gen I.

The top view of the WCFB four-barrel carburetor, which is the predecessor of the AFB. Note that inside the round air horn, it has a square shape.

CARTER CARBURETOR HISTORY

The Carter Carburetor company was founded by Will Carter in 1909. He patented his Model C in 1910. Although the Dodge Brothers car company was an early customer, Carter focused on the aftermarket during their first years in business. Their first OEM production orders came from Chevrolet in 1925 and Chrysler in 1928.

While there have been many carburetor manufacturers over the years, there have only been four major players in OEM production—Carter, Holley, Rochester (GM) and Ford/Autolite. The Carter carburetors that will be discussed are called the AFB, the AVS and the TQ, or Thermo-Quad.

The last of the production Carter carburetors was built in the mid-1980s. Typically once the production of a part is stopped, it becomes very difficult to use that part because you can't get new parts or service parts. This is probably even more important on a carburetor because carbs do wear out. In the 1990s, Edelbrock purchased the Carter Carburetor tooling and has brought back the AFB, and more recently the AVS, and all the parts required to tune it for almost any application.

Carter Model Overview

BBD—The BBD is a two-barrel Carter carburetor that first appeared on production models in 1952. They came in 1 1/4" and 1 1/2" diameters (of the throttle bores). They were produced into the early 1980s. The letters stand for "Ball and Ball" and "dual." There was also a BBS, which was the single or one-barrel design. The basic design for the BBD was purchased from the Ball Brothers by Carter Carburetor.

WCFB—The WCFB was the first Carter four-barrel carburetor and was introduced in 1952. The letters stand for "Will Carter Four Barrel." The early models flowed around 385 cfm. The carburetor was made from three castings—the throttle body (bottom) was made from cast iron, the main body (middle) was made of zinc and the air horn (top) was made of aluminum. The WCFB is heavy! It was first used on Buick straight-eight

CARTER HISTORY, BASIC THEORY & IDENTIFICATION

This is the 1966–'71 production 426 Street Hemi that uses two four-barrel AFB carbs mounted inline underneath the large round air cleaner top. It was the last production use of the AFB carburetors in 1971 model year.

The AVS carbs, Edelbrock version shown, were only in production for six years. There were replaced by the solid-fuel Thermo-Quads.

engines through 1965. It can be identified by the square-in-circle style of the WCFB's air horn (top).

AFB—The AFB (aluminum four-barrel) was introduced in 1957 and is arguably the most famous Carter model. The AFB was used in many dual four-barrel production applications, especially in the late 1950s and early 1960s, and on the '66–'71 426 Street Hemi engines and the early '60s 348/409/427 Chevys. Both engines used eight-barrel inline induction systems.

The original early models flowed 450–625 cfm. The AFB four-barrel carb was truly universal, installed on nearly every U.S. production engine from the late '50s through 1971. The last production application was as a dual four-barrel inline system on the 1971 426 Street Hemi engine. As a high-performance carb, it was very much a key component of the muscle car era. The AFB uses a counterweighted air velocity valve located just above the secondary throttle blades.

AVS—The AVS (air valve secondary) was introduced in 1966. It was used on production engines through the 1971 models except for the Hemi. While they look similar to the AFB, the AVS features a spring-loaded air valve located over the secondary bores at the top of the air horn. Additionally there are no secondary booster venturis (clusters). When new emissions rules were mandated in 1972, the AVS models were replaced by the solid-fuel Thermo-Quads. The production AVS carburetors used nozzle bars in the secondaries to replace the secondary venturis. This makes them unique from the Edelbrock AVS, which uses secondary venturis similar to the AFB. The AVS is much easier to adjust to specific engine packages and driving conditions than the AFB because the

A 3/4 front view of the production AVS carburetor.

secondary air door is adjusted by a spring. Between the metering rods and the adjustable air door, these carbs can be adjusted for most conditions without taking the actual carburetor apart.

Thermo-Quad—The Thermo-Quad (TQ), was introduced in 1971 on the Chrysler 340 high-performance engines. However, the 1972-and-later Thermo-Quad is a much more popular model. Thermo-Quads were used in production through the mid-1980s. The Thermo-Quad is perhaps the most unique carburetor built for production car use. Where most carbs are made of aluminum castings or zinc castings, the center section (main body) of the Thermo-Quad is made of black, molded phenolic resin (plastic). The top (air horn) and bottom are made of aluminum. The phenolic resin acts as a very effective heat insulator and the fuel in the carb is kept cooler (approximately 20

The 3/4 front view of the 1972-and-newer Thermo-Quad. Note the two vertical pipes at the rear of the air horn, which identify it as the solid fuel, 1972-and-newer version. The '71 version has one large vertical pipe.

The front view of the Edelbrock AFB four-barrel carburetor. They are easily identified by the Edelbrock logo located in the center of the carb just below the air cleaner support ring.

The production AFB and AVS carburetors up through 1971 used a mechanical bowl vent that was open to the atmosphere. The 1972-and-newer production Thermo-Quad carburetors were solid fuel designs and did not vent to atmosphere. Vapors were collected and vented to a charcoal canister through this fitting. The Y-link to the lower right operates the actual internal vent.

degrees F) than in all-metal carburetor designs. It also has very small primary throttle bores (1 3/8" and 1 1/2") and very large secondary throttle bores (2 1/4"). This is an air-bleed carb in that it vents to the atmosphere similar to all the earlier AFB and AVS carbs. The '72 Thermo-Quad was used on all Chrysler V8s and did not vent to atmosphere. Instead it has a hose connection to a charcoal canister, typically located in the engine compartment on the right side. The '71 Thermo-Quad carbs were superseded by the '72 carbs.

The 9000 Series—The 9000 Series AFB carburetors were introduced by Carter in the mid-1970s. They were designed as bolt-on replacement carbs for GM, Chrysler and Ford engines. They featured electric choke, emissions connections and OEM throttle linkage compatibility and were recalibrated to meet new emissions standards. The Carter Competition Series is also based on this carb model. Carter was purchased by Federal-Mogul in about 1986 and ceased making carburetors shortly thereafter.

Edelbrock—There are two styles of Edelbrock carburetors—the Performer Series are similar to the AFB and the newer Thunder Series are similar to the AVS. Both of these carbs are derived from the original 9000 series carbs. They also feature a dual-bolt pattern mounting flange and are assembled using Torx screws.

Edelbrock offers a rebuild kit for both of their models. Rebuild kits for the other three original Carter production carbs should be available from your local auto parts store. Edelbrock offers individual components like gaskets, accelerator pumps, and linkage parts, which can be handy if you have just one or two missing or broken pieces.

Edelbrock offers special versions of the AFB and AVS carbs that are specially designed to be emissions-legal in all fifty states, and others are designed for marine use.

Edelbrock also has a calibration for using a 700–800 cfm single four-barrel on a supercharged engine.

Dual Four-Barrel Systems—There are two basic types of eight-barrel carburetor systems and perhaps the first one was the inline system which has two four-barrel carburetors sitting on top of the engine, one behind the other. While many of the eight-barrel carburetor systems date back to the 354–392 old-style Hemis, now called Gen I, these were originally based on the WCFB carburetors. The eight-barrel inline package was probably more popular on the early wedge engines, especially in the early 1960s. However, the eight-barrel reached its production peak when it was used on the

CARTER HISTORY, BASIC THEORY & IDENTIFICATION

There are many styles of linkage now available for the eight-barrel inline carburetor setup. This is a mechanical setup that opens both carburetors at the same time—probably for racing.

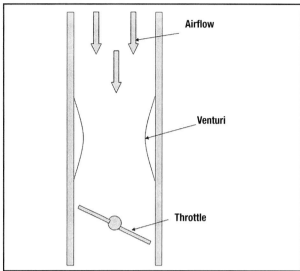

The basic principle is pretty straightforward with the vertical throttle bore, the throttle blade at the bottom and the venturi. The top of the throttle bore is open to atmosphere so the air can flow in. The venturi is smaller than the throttle bore and helps speed up the airflow which helps draw the fuel out of the carburetor.

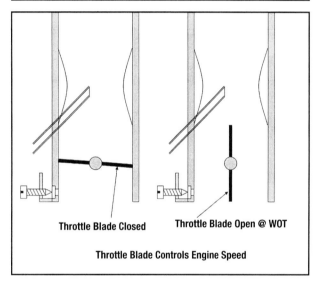

The throttle blade is located at the bottom of the vertical throttle bore. The basic position of the throttle blade will control the engine speed and airflow. With the throttle blade closed, the engine is in the idle configuration as on left. With the throttle blade open (vertical) the engine is in wide-open throttle (WOT) mode, as on the right

1966–'71 426 Street Hemi engines (Gen II). The unique aspect of the eight-barrel inline package is that the two carbs are different—the rear carb has a choke while the front carb does not. Early production versions were used on the 348–409 Chevies and the 343hp, 383 and 385hp 413 Dodge/Plymouth in 1962.

The very early eight-barrel system was called a cross-ram, and had very long runners and one carb on each side of the engine. These long-ram systems were used on late '50s and early '60s Chrysler vehicles. They were probably the first tuned manifolds used in production. Each carburetor actually sat outside of the engine's valve cover. While these packages are visibly impressive, they tune at too low of an rpm to be competitive with today's high-flow, high-rpm engine hardware. In 1962, a much shorter runner, one-piece cross-ram was used on the 413 Max Wedge engines. It was also used on the '63–'64 426 versions. This is a one-piece casting and was the most popular of the production-based cross-ram packages. A similar eight-barrel cross-ram was used on the 426 Race Hemi engines (Gen II) in '64–'65 and '68 but these sold a few hundred packages while the Max Wedges sold a few thousand. The cross-ram system uses two carbs that are the same. The AFB carburetor also reached its peak with the '64 package and the use of the 3705 AFB, which featured the largest throttle bore/venturis and highest airflow of the production carburetors.

Solid Fuel—Technically, this term means that the fuel metering system feeds the primary and secondary discharge nozzles with a solid, continuous stream of fuel that mixes with air AFTER it leaves the nozzles. This solid fuel aspect makes these carburetors unique from the '71 Thermo-Quad (TQ) and the AVS and AFB models built prior to '72, which use the conventional air-bleed metering systems that introduce the air into the fuel stream before the fuel discharges from the nozzles. Perhaps an easier explanation is that the 1971-and-earlier carbs all vent to the atmosphere and have a bowl vent, which is part of the carb linkage, usually close to the accelerator pump. The 1972-and-newer, solid-fuel TQs are sealed and use a hose to vent the gases to a charcoal canister, which is connected back to the fuel tank. Therefore, no gas fumes escape into the atmosphere.

The float sits inside the bowls on the carburetor. On Carter/Edelbrock carbs there are two floats in each carb. The float is hollow or made of very light material. The fuel pushes up on the float, and this force in turn pushes on the needle. When the level of fuel gets high enough, enough force is generated to shut off the flow of fuel through the needle and seat.

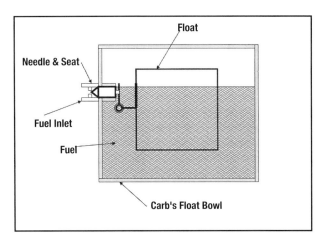

BASIC CARBURETOR THEORY & OPERATION

The carburetor's basic function is to mix the fuel with the air being delivered to the engine's cylinders. The scientific discussion of the mixing process and the resulting flow of air/fuel mixture into the cylinders and ignition requires some special terms that need to be defined.

Basic Operation

When na engine cylinder's piston moves down the cylinder bore on the intake stroke, it draws air into the cylinder from the intake manifold, creating a vacuum. The vacuum effect from the manifold then draws air through the carburetor bores. As air flows by the venturi, fuel is drawn from the carb's float bowl and is mixed with air. The resulting air/fuel mixture passes into the intake manifold and

With the throttle blades closed, the engine depends upon the idle system that is built into the carburetor. The key part of this system is the idle mixture screw, which is located below or after the throttle blade. This screw can help set the engine's idle by adjusting how much fuel is sprayed into the system.

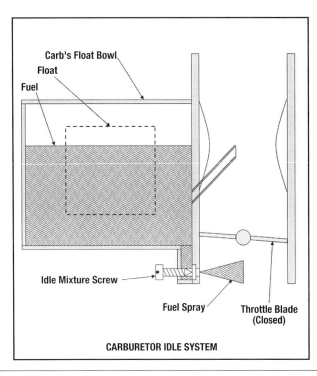

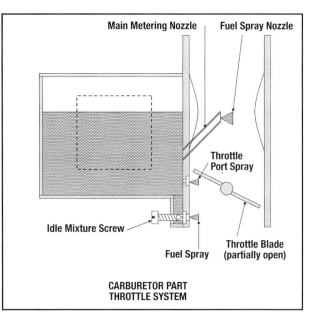

The part throttle fuel system has three separate parts. At part throttle the blades are only slightly open and in this position there is only a small amount of air that passes through the throttle bore. You have the idle spray and some spray from the transfer port, plus a small amount of fuel from the main metering system, but this is limited because the airflow is limited in this throttle position.

onto the cylinders for combustion.

Carb Size—A carburetor can be rated by how big the throttle bore is. A one-barrel carburetor has one bore flowing air, a two-barrel has two bores and a four-barrel has four bores. While the left and right bores tend to be the same size, the fronts in a four-barrel may be equal to or smaller than the rears. Many production carbs were rated as either 1 1/4" or 1 1/2" bore two-barrel, or a 1 3/4" bore four-barrel. The aftermarket seems to prefer to rate the carbs on total airflow, so you have a 600 cfm four-barrel or an 800 cfm four-barrel.

CFM—Cubic feet per minute (cfm), is the standard unit of airflow measurement. It is used to rate carburetors, intake manifolds and manifold/head combinations. Cfm is measured on a flow bench. A flow bench is expensive and therefore it is not really a home shop tool. Typically, the flow bench numbers will be tested and given to you by your machine shop/engine builder/porting service. While flow benches were only available at the OEMs in the 1960s, today they have become common at almost all shops that deal with engines and cylinder heads. There are three cfm ratings that are mainly discussed—carb, cylinder head/intake manifold and engine's cfm. The first two are measured while the last one is a calculation.

Engine Airflow—An engine's airflow is usually measured in cubic feet per minute. There is an

CARTER HISTORY, BASIC THEORY & IDENTIFICATION

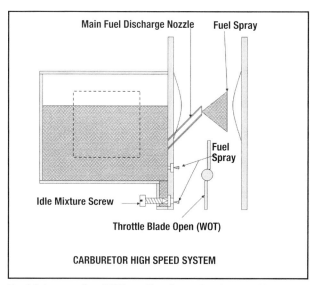

For high-speed or WOT applications, the throttle blade goes vertical and opens fully. This is the maximum airflow position, and while the idle and transfer slots or parts still work, they are a very small part of the whole, which is mainly coming from the main fuel discharge nozzle located near the throttle bore's venturi.

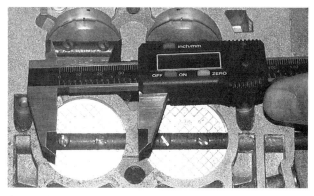

Carter OEM production carburetors were not cfm rated, but the Edelbrock carbs are. But it can be helpful to actually measure your carburetor's throttle bore sizes.

Since the venturis are located inside the carburetor they are not easy to see with the carb assembled. With the throttles held open, you can see the venturis above the vertical throttle blades and next to the cluster/nozzle rings.

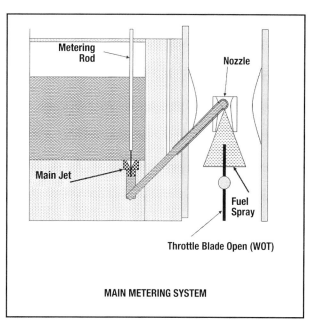

equation that relates the carb size to the engine displacement and engine rpm. It is discussed in Chapter 3, starting on page 31.

Fuel—The carburetor's job is to meter and control the flow of air and to properly mix the fuel into the air stream. Most of the focus has always been on the cfm rating of the carb, but not much consideration has gone into fuel metering. In the first twenty years of the 426 Hemi, alcohol was only used with fuel injection in dragsters. Since the mid-1990s, alcohol fuel has become more popular in bracket racing and is available on the street as E85 fuel (15% gas and 85% alcohol). The type of fuel affects almost every passage in a carburetor except for the throttle bores. Alcohol fuel, specifically methanol, needs to be twice as rich as gasoline, so the jets must be much larger. Edelbrock offers special carbs just for use with alcohol-based

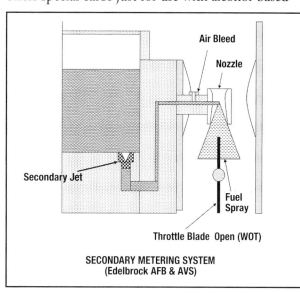

The main metering system delivers fuel to the nozzles or cluster located in the throttle bore's venturi. The fuel in the float bowl passes through the main jet, which has the metering rod sitting in it. The resulting area of the two parts controls the actual fuel flow. Then passages allow the fuel to get to the nozzle and be pulled out by the lower pressure in the venturi with the throttles open. This is the system that is used in the primaries of the carburetor.

The secondary metering system as used on the AFB and Edelbrock AVS carburetors does not have a metering rod sitting in the main jet. The jet opening itself is the only fuel control. There is an air bleed in the cluster nozzle that is important, but isn't adjusted.

The secondary metering system used on the Thermo-Quad is different from the AFB style. The TQ uses a tall secondary jet and two nozzle bars in place of the clusters/nozzles that are used in the AFB. The TQ is shown but the production AVS also uses nozzle bars in the secondary so it would look similar it just doesn't use the tall jet.

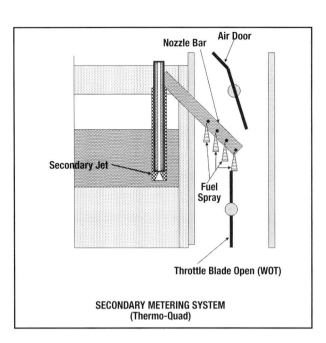

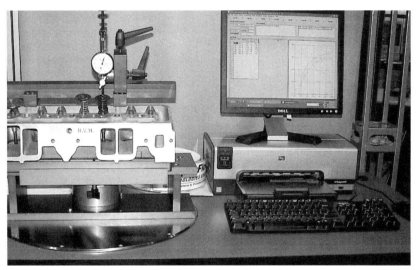

Information on cylinder head flow comes from a flow bench. It flows one cylinder at a time and one port of that cylinder at a time. This one is set up for the intake. The head bolts to a fixture that simulates the cylinder bore size. The machine's software program controls much of the numbers but the result is a graph of the airflow at various valve lifts. The curve is shown in the upper right of the photo. The dial indicator is used to precisely measure valve lift.

fuels. The other type of alcohol fuel, ethanol, only needs to be about 35–40% richer than gas. The E85 fuel also needs to be richer—about 25–30%. If you run any of these fuels, you need to recalibrate the carb by changing jets and metering rods.

Note: Gasahol is a blend of not more than 10% ethanol and gasoline. At the maximum 10% mix, the same calibration can be used as with straight gasoline. Fuel is covered in greater detail in Chapter 6.

Cylinder Head/Intake Manifold Airflow—The cylinder head cfm is probably the most popular airflow number discussed. For example, your head could flow 250 cfm, which actually means that one intake port flows 250 cfm at peak—typically at max valve lifts like 0.600", 0.700" or 0.800". Typical intake ports (wedge-style) flow from 175 cfm to 275 cfm, while a stock Hemi port might flow close to 300 cfm. Fully ported intake ports can flow 350 cfm to over 400 cfm. Racing Hemi and Pro Stock engine ports are much higher and larger, with much larger intake valves. These numbers are all peak flow numbers, and what you really want is the flow curve, the amounts of airflow at the points in between, in addition to the peak numbers. There are also flow numbers for the exhaust side, like 200 cfm, but they are always less than the intake (about 60–80% of the intake flow number) and therefore are not as popular to discuss.

Carburetor Airflow—The carburetor's airflow number is typically given to you by the manufacturer. Carbs are rated at 500, 600, 750 and 800 cfm. The OEMs did not flow-rate the production carbs by cfm but tended to use the throttle-bore size as the carb comparison number. The carb airflow number is usually very close to the engine's airflow number calculated by the equation in Chapter 3. Since this is how the carb model should be selected, this should always be true.

Perhaps the most confusing aspect of cfm directly relates to the two-barrel engines and the 340 and 440 engines, which use three two-barrel carbs for a total of six barrels. More specifically, this confusion stems from the airflow ratings of the two-barrel carbs. For example, the center carb of a six-barrel system is rated by Holley (the manufacturer) at 350 cfm and the end carbs are rated at 500 cfm. That would make the complete package flow 1350 cfm (350 + 500 + 500 = 1350). The real airflow might be closer to 1000 cfm. Let me explain—the standard carburetor package in the performance world is a four-barrel carb. Many years ago, the leading carb manufacturer, Holley Carburetors, defined the rating system for four-barrel carbs as 1.5" of mercury (pressure drop). Carter, the other major four-barrel carb manufacturer at the time, accepted this standard and Holley was the largest aftermarket manufacturer. No one even says 1.5" when they discuss four-barrel carb ratings today. Just the cfm number. The trick is that the carb manufacturers also specified that two-barrel carbs be rated at 3" of mercury. A carb will flow more air when rated at 3" than it does at 1.5". These 350 and 500 cfm two-barrel carbs are basic 1 3/4" throttle bore designs and are the only two-barrel carbs used in racing. The BBD two-barrel carbs have 1 1/2" throttle bores and are too small to be competitive.

BSFC—Brake specific fuel consumption. BSFC is the ratio of fuel consumed (in pounds per hour) to the brake horsepower that's measured on a dyno. For a typical gasoline engine, this ratio is about 0.50. On the other hand a supercharged engine would be closer to 0.60. If you use methanol for fuel, then the BSFC is doubled. For example, how much gasoline does your 600 hp Hemi need? Answer: 600 x 0.50 = 300 lbs/hr.

Note: To convert lbs/hr to gallons per hour, divide by 6 (the weight of one gallon of fuel). So 300 divided by 6 = 50 gph. So if you have an electric fuel pump that pumps 70 gph at 7 psi head pressure, you should be covered.

Free Flow—Fuel pumps are rated in gph, or gallons per hour. This gph rating is generally given at a specific pressure like 7 psi. Free flow is with a pressure of zero. If you disconnect the fuel line before the carburetor and run the pumps to check the amount delivered in 20 seconds (standard pump check), this open fuel line represents free flow. Fuel pumps will pump more fuel at free flow than they do at a specific pressure. The 5 to 7 psi numbers are typical of carburetor needle and seat assembly's desired pressure for best operation, and therefore makes a good rating level.

On electric fuel pumps, the other ratings trick is voltage. Most cars today use a 12-volt system. However some pumps are rated at 14-volts. The electric pump will pump more fuel at 14-volts than at 12-volts.

Air/Fuel Ratio—The ratio of air mixed with fuel is very important to the operation of the engine, and it is the primary job of the carburetor. The resulting ratio of the air to fuel is defined as the air/fuel ratio. This ratio is based on weight (or mass). Therefore if 14 pounds of air are combined with 1 pound of fuel, the A/F ratio is 14-to-1 or more commonly A/F = 14. This number happens to be the chemically ideal or correct ratio of air to fuel for complete combustion. It is known as the *stoichiometric ratio*. It is based on the weight of incoming air and flammable gas/vapor (fuel) at which complete combustion will take place. This means that there would be no excess fuel or oxygen (key part of air) after combustion. The actual stoichiometric ratio (for gasoline) is 14:1. It is the number that all tuners strive for.

"Rich" is a term used to describe a cylinder that has more fuel than is needed for ideal or complete combustion. It means that the A/F ratio is less than 14.

An engine that is running "lean" is a condition of having too much air. That means that the A/F ratio is more than 14.

Engines can run on A/F ratios of 5 to 25, but the

AIR/FUEL RATIO

A/F ratio	Typical Characteristics
6–9	Extremely rich, w/black smoke and low power
10–11	Very rich
12–13	Rich, but the best power ratio for wide-open throttle conditions
14	Stoichiometric, no excess fuel or oxygen/air. Good A/F ratio for part throttle cruise and light to moderate acceleration.
15–16	Lean, but the best economy A/F ratio, borderline for part-throttle drivability
17–18	Very lean, usually the limit of drivability before severe engine damage

Note: The engine will run in both extremes of lean or rich A/F ratio. The rich extreme makes black smoke while the lean extreme can damage the engine by overheating, scuffing pistons or rings or by blowing head gaskets.

The numbers stamped on the baseplate flange as shown here at the lower center—right side as installed on the engine—are used for identification. Generally, it is the four digits of the carb number—1407 as shown.

normal range is much more narrow. An A/F ratio of about 12.5 is used for wide-open throttle (WOT) and 14.5–15.5 for part throttle or cruising. An intermediate A/F number of 13–14 is typically used for midrange power, and acceleration without being at WOT.

CARBURETOR ID
Model Numbers

Each production manufacturer (OEM) will give a specific carburetor its own part number. These part numbers usually have eight digits or more. On the other hand, a specific carb manufacturer like Carter will give the carburetor a second number for identification. These numbers tend to be about four digits long and are typically stamped into the casting of the carb itself. Most of the AFB and AVS carbs have the number stamped on the right-front boss that is used to attach the carb to the intake manifold. Production carbs also have a metal tag attached to one of the assembly screws on the right

Production carburetors also use a stamped tag that is held on by one of the attaching screws. They are typically located on the right front. These tags are not used on Edelbrock carbs.

The wedge-head, dual inline systems tended to be single plane intakes. Note choke is on right rear.

The '72-and-newer Thermo-Quads have two vertical pipes at the rear. The shorter pipe is in the center and is actually located at the front of the air horn.

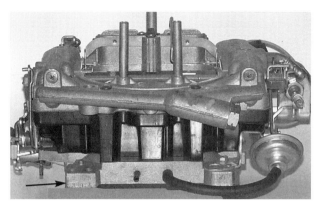

The Thermo-Quad numbers are stamped in the baseplate flange but are located on the rear edge rather than the front. Linkage side or lower left in photo.

The '71 Thermo-Quad was the first TQ but only lasted one year. They can be identified by the large, single vertical pipe in the center-rear.

The Edelbrock AFBs are brand-new carburetors and look just like the originals except for some very minor details not easily seen in a photograph.

front corner. These tags are often lost over time so the stamped numbers tend to be more reliable. On Thermo-Quad carbs, they tend to stamp the number on the left rear boss. In both the AFB/AVS and TQ examples, the stamped number is on the vertical surface so you must look at the carb from the side to see the number.

The AFB, AVS or TQ lettering precedes the four-digit number and is not usually stamped on the carb.

PRODUCTION CARB NUMBERS

AFB Dual Four-Barrel Systems

Car	Year	HP	Displacement	Carb Numbers
Buick	'64–'65	360	425	AFB-3645, 3634, 3646
Chevy	'56	225	265	WCFB-2626, 2627*
	'57	270	283	WCFB-2626, 2627, 2419, 2362*
	'57	245	283	WCFB-2626, 2627*
	'58	245	283	WCFB-2613, 2614, 2626, 2627*
	'58	270	283	WCFB-2613, 2614, 2626, 2627*
	'59	245	283	WCFB-2626, 2627*
	'59	270	283	WCFB-2613, 2614*
	'60–'61	245	283	WCFB-2626, 2627*
	'60–'61	270	283	WCFB-2613, 2614*
	'61–'62	409	409	AFB-3361, 3362
	'63	425	409	AFB-3361, 3362
	'63	430	427	AFB-3361, 3362
	'64	425	409	AFB-3361, 3362
Chrysler	'60	330	383	AFB-2903 (2)
	'60	380	413	AFB-3084 (2)
	'61	375	413	AFB-2903 (2)
	'61	400	413	AFB-3084 (2)
	'62	400+	413	AFB-3084 (2)
	'62	380	413	AFB-3258, 3259
	'63–'64	390	413	AFB-3505 (2)
Dodge/Plymouth	'60	310	361	AFB-2903 (2)
	'60–'61	330	383	AFB-2903 (2), 2790, 2791
	'61	340	383	AFB-3084 (2)
	'61	375	413	AFB-3084 (2)
	'62	343	383	AFB-2790, 2791, 3258, 3259
	'62	385	413	AFB-2790, 2791, 3258, 3259
	'62	394	413	AFB-2790, 2791, 3258, 3259
	'62	410	413	AFB-3447, 9626
	'62	420	413	AFB-3447, 9626
	'63	415	426	AFB-3447, 9626
	'63	425	426	AFB-3447, 9626
	'64	415	426	AFB-3705
	'64	425	426	AFB-3705
	'66	425	426	AFB-4139, 4140
	'67	425	426	AFB-4139, 4140, 4342, 4343, 4619, 4620
	'68	425	426	AFB-4430, 4431, 4620, 4430, 4432
	'69	425	426	AFB-4619, 4620, 4621
	'70	425	426	AFB-4742, 4971, 4619, 4620, 4745, 4969
	'71	425	426	AFB-4971, 4969, 4970
Ford/Shelby	'63	289	289	AFB-2190, 2191
	'64–'65	300	289	AFB-3258, 3259
Pontiac	'62	405	421	AFB-3433, 3435

*WCFB numbers listed for reference only.

AVS Four-Barrel Systems

Car	Year	HP	Displacement	Carb Numbers
Chevy	'66	300	327	AVS-4028
Dodge/Plymouth/Chrysler	'68	275	340	AVS-4424, 4425, 4611, 4612, 4639
	'68	335	383	AVS-4426, 4401
	'68	375	440	AVS-4428, 4429, 4617, 4618, 4640
	'69	275	340	AVS-4611, 4612
	'69	300	383	AVS-4613, 4616, 4682
	'69	335	383	AVS-4613, 4616, 4682
	'69	375	440	AVS-4617, 4618, 4640
	'70	275	340	AVS-4933, 4936, 4934, 4937
	'70	330	383	AVS-4736, 4734, 4732
	'70	375	440	AVS-4737, 4739, 4738, 4741
	'71	335	440	AVS-4966
	'71	370	440	AVS-4967, 4968

AFB Four-Barrel Systems

Car	Year	HP	Displacement	Carb Numbers
AMC	'66	225	290 V8	AFB-4250, 4252
	'67	280	343	AFB-4216
	'68	226	290	AFB-4467, 4622, 4585
	'68	280	343	AFB-4469, 4624, 4468, 4623
	'68	315	390	AFB-4583, 4584
	'69	225	290	AFB-4660
	'69	280	343	AFB-4662
	'69	315	390	AFB-4664
Buick	'57–'58	300	364	AFB-2507
	'59	325	401	AFB-2877, 2840
	'60	300	364	AFB-2981
	'60	325	401	AFB-2982
	'61	300	364	AFB-3089
	'63	325	401	AFB-3503
	'63	340	425	AFB-3579
	'64	325	401	AFB-3635
	'64	340	425	AFB-3635
	'65	250	300	AFB-3827
	'65	325+	401-425	AFB-3922
	'66	260	340	AFB-4056
	'66	325	401	AFB-4060, 4054, 4053, 4059
Cadillac	'59	325	390	AFB-2814, 2815
	'60	325	390	AFB-2814, 2951
	'61	325	390	AFB-3177, 3178
	'62	325	390	AFB-3351, 3352
	'63	325	390	AFB-3480, 3481
	'64	340	429	AFB-3655, 3656
	'65	340	429	AFB-3903
	'66	340	429	AFB-4168, 4169, 4171

(AFB Four-Barrel Systems, continued)

Car	Year	HP	Displacement	Carb Numbers
Chevy	'59–'60	305	348	AFB-2897
	'59–'60	320	348	AFB-2897, 3012
	'61	305	348	AFB-3221
	'61	360	409	AFB-3270
	'61	409	409	AFB-3361, 3362
	'62	305	327	AFB-3269
	'62	340	327	AFB-3269
	'62	380	409	AFB-3345
	'62	409	409	AFB-3361, 3362
	'63	300+	327	AFB-3461
	'63	400	409	AFB-3499, 3783
	'64	300	327	AFB-3461, 3460, 3459, 3462
	'64	400	409	AFB-3499
	'65	400	409	AFB-3738
Chrysler	'57		354	AFB-2448, 2686
	'57		354–392	AFB-2448, 2686
	'58		354–392	AFB-2650, 2651, 2805, 2806
	'59	305	361	AFB-2773, 2787, 2794
	'59	350	413	AFB-2797
Dodge/Plymouth	'58		361	AFB-2642, 2652, 2653
	'59		361	AFB-2773, 2787, 2794
	'60	255	318	AFB-2948, 2991
	'60	325	383	AFB-2968
	'60–'61	148	170	AFB-3083
	'61	196	225	AFB-3083
	'61	260	318	AFB-3103
	'61	305	361	AFB-3105, 3106
	'61	325	383	AFB-2968, 3113
	'61	350	413	AFB-3108, 3113
	'62	260	318	AFB-3247, 3249
	'62	305	361	AFB-3252, 3253, 3257
	'63	330	383	AFB-3437
	'64	330	383	AFB-3611
	'64	360	413	AFB-3615
	'64	365	426W	AFB-3611
	'65	235	273	AFB-3853, 3854
	'65	315	383	AFB-3855, 3856
	'65	330	383	AFB-3859, 3860
	'65	340	413	AFB-3858
	'65	360+	413–426	AFB-3859, 3860
	'66	235	273	AFB-4119, 4121, 4120, 4122
	'66	325	383	AFB-4130, 4132
	'66	350	440	AFB-4130, 4136
	'67	235	273	AFB-4294
	'67	280	383	AFB-4309
	'67	325	383	AFB-4298
	'67	350	440	AFB-3112
	'67	375	440	AFB-4326

(continued on next page)

(AFB Four-Barrel Systems, continued)

Car	Year	HP	Displacement	Carb Numbers
Ford-Mercury	'57–'59	300	352	AFB-2441, 2640
	'60	350	430	AFB-2992
Lincoln	'59	375	430	AFB-2853
	'60	300	352	AFB-2292
	'63–'64	315	430	AFB-3521, 3522
Pontiac	'57–'59	270	347	AFB-2506
	'58	255	370	AFB-2740, 2751, 2767
	'58–'59	260	389	AFB-2819, 2820
	'58	315	370	AFB-2768, 2767
	'60	235	389	AFB-2975
	'60	333	389	AFB-2976
	'61–'62	235	389	AFB-3123
	'61	287	389	AFB-3125
	'61	303	389	AFB-3124
	'61	333	389	AFB-3123, 3125
	'62	303	389	AFB-3300, 3326
	'63	280	326	AFB-3477, 3502
	'63	235	389	AFB-3479, 3423, 3426, 3428
	'63	353	421	AFB-3574, 3544
	'64	280	326	AFB-3686, 3687
	'64	303	389	AFB-3648, 3623, 3629, 3647
	'64	320	421	AFB-3650, 3651
	'64	325	389	AFB-3647, 3648
	'65	285	326	AFB-3899
	'65	325	389	AFB-3896
	'65	333	389	AFB-3895
	'65	338	421	AFB-3895, 3898
	'66	285	326	AFB-4035, 4036
	'66	325	389	AFB-4034, 4033
	'66	338	421	AFB-4037, 4033
	'67	285	326	AFB-4243, 4246
	'67	350	400	AFB 4242, 4243
Studebaker	'63–'64	225	289	AFB-3588, 3540
	'63–'64	195	259	AFB-3589, 3506, 3507

Thermo-Quad Four-Barrel Systems

Car	Year	HP	Displacement	Carb Numbers
Dodge/Plymouth/Chrysler	'71	275	340	TQ-4972, 4973 **
	'72	240	340	TQ-6138, 6139
	'72	255	400	TQ-6140, 6165, 6090, 6166
	'72	265	400	TQ-6140, 6165, 6090, 6166
	'73	240	340	TQ-6318, 6339, 6319, 6340
	'73	220	440	TQ-6322, 6410
	'73	260	440	TQ-6320, 6341, 6321, 6342
	'73	280	440	TQ-6324, 6311
	'74	200	360	TQ-9013, 6453, 6488
	'74	205	400	TQ-6489, 6459, 9014
	'74	230	440	TQ-9015, 9023, 6024
	'74	240	400	TQ-6456
	'74	250	400	TQ-6456
	'74	245	360	TQ-6452, 6454, 6453, 6455
	'74	250	440	TQ-9015, 6023, 9016, 9024, 6460, 6461, 6462, 6463
	'74	275	440	TQ-9015, 6023, 9016, 9024, 6460, 6461, 6462, 6463
	'75–'76	190+	360	TQ-9002, 9004, 9055, 9076, 9093
	'75–'76	185+	400	TQ-9008, 9046, 9050, 9053, 9056, 9057, 9064, 9074, 9097
	'75–'76	210+	440	TQ-9012, 9052, 9073, 9009, 9010, 9051, 9058, 9059, 9095, 6545
	'76	250+	440	TQ-9011, 9094, 9065
	'77	160+	360	TQ-9076, 9073, 9093, 9104, 9115
	'77–'78	190	400	TQ-9077, 9102, 9103
	'77–'78	195+	440	TQ-9078, 9101, 9114, 9119, 9080, 9081
	'78–'79	150+	318	TQ-9123, 9137, 9121, 9147, 9195, 9245, 9202, 9256
	'78–'79	160+	360	TQ-9134, 9104, 9197, 9250, 9196, 9202, 9246
	'80	150+	318	TQ-9232, 9234, 9295, 9243, 9306, 9320
	'80	185+	360	TQ-9244, 9305, 9236
	'81–'82	165+	318	TQ-9364, 9372, 9373
	'83	150+	318	TQ-9364, 9374, 9388
	'84	165	318	TQ-9389, 9390
International	'74–'76		345	TQ-6551, 6590, 6592
	'74–'76		392	TQ-9027, 9028

** Superseded by TQ-6138, 6139

Note 1: The plus (+) sign added to the end of the horsepower number in the HP column indicates that there are more than one horsepower rating at this displacement.
Note 2: In the charts on the AFB, AVS and TQ production carburetors, the majority of the truck engines/models were not included. Only passenger cars.
Note 3: In the charts on the AFB, AVS and TQ production carburetors, there are many tricks because often times carburetors were dual-sourced, which means another brand may have also been used on specific engines. There were several union strikes over the years that resulted in other manufacturer's carburetors being used in place of the Carter Carburetor listed above. But it also worked in reverse—Carter carbs were used in place of other styles of carbs. In many cases, the situation only lasted for a few months but it is very difficult to track.

EDELBROCK AFB AND AVS SYSTEMS
(Aftermarket)

Model	CFM	Choke	Carb Numbers
Performer AFB	600	electric	1400 (50-state legal)
	500	electric	1403
	500	manual	1404
	600	manual	1405
	600	electric	1406
	750	manual	1407
	600	electric	1409 (marine)
	750	electric	1410 (marine)
	750	electric	1411
	800	manual	1412
	800	electric	1413

Models 1403 and 1404 can be used on small displacement engines or dual-quad applications
Edelbrock offers models 1405(4), 1406(4) in EnduraShine, a polished finish (the 4 denotes the finish)

Model	CFM	Choke	Carb Numbers
Thunder AVS	500	electric	1801
	500	manual	1802
	500	electric	1803
	500	manual	1804
	650	manual	1805
	650	electric	1806
	800	manual	1812
	800	electric	1813
	650	manual	1825
	650	electric	1826

Models 1803 and 1804 are calibrated for dual-quad use
Edelbrock offers models 1801(4), 1802(4), 1803(4), 1804(4) in EnduraShine, a polished finish (the 4 denotes the finish)

Note 1: With carburetor tuning, it is a very good idea to keep track of your changes, i.e. write it down. This is best done on a run sheet or log that lists all the changes made to the carb and the results. The engine is an important part of the team so the general specifications for the engine should be listed first. This notebook/log can be a part of your engine-build book or separate. Then list the carburetor model number and general specs, throttle bore or cfm size. Then make a chart that lists the data as follows: date, primary jet, secondary jet, metering rod, spring, accelerator pump link, accelerator pump shooter size, needle and seat, air cleaner, fresh air and then the results. In drag racing the results would be based on speed and ET (elapsed time). For other applications, the results might be based on fuel economy or drivability.

Chapter 2
Carter/Edelbrock Components & Tech Specs

Left side of the production AFB carburetor. Note the bowl vent in the center and the accelerator linkage to the left and compare the somewhat simple secondary linkage to the complicated linkage used on the Thermo-Quad shown in the photo on page 121.

In this chapter I will only discuss the basic four-barrel Carter AFB, AVS, Thermo-Quad and Edelbrock Performer Series and Thunder Series carbs. In general, the production AFB and AVS carburetors look very similar, while the Thermo-Quad looks unique in several ways. The Edelbrock carbs look similar to the production versions of the AFB and AVS except for the Edelbrock logo in the center front part of the carb just below the air cleaner seat ring.

Carb Components

The carburetor is not a large part relative to something like a cylinder head, but it has quite a few components. The jets and metering rods are changed for tuning the carburetor to your specific application or to any engine package. Most of these pieces do not require replacement unless they are damaged. Typically, if a carburetor receives that much damage, it should probably be replaced since a wide selection of new carbs are readily available.

Housings—The AFB and AVS, along with the Edelbrock versions, use two housings, a bottom and a top (air horn). Both are made of aluminum. The Thermo-Quad is a three-piece construction with the top (air horn) and bottom being made of aluminum while the centerpiece is made from black, molded phenolic resin (plastic). If a housing is seriously damaged, it should be replaced, which would likely mean a new carburetor assembly. There are 8–10 screws that hold these parts together. The key is to use the correct gasket between these parts.

Note: The production AVS carbs use nozzle bars in the secondaries while the Edelbrock AVS versions use two secondary venturi similar to the AFB models.

Carb Flange—There are several different bolt patterns used on the carb flange and two different shapes. The production carbs tended to have only one pattern machined into the carb's attaching flange (usually 4 holes but

The top view of the production AVS. The primaries are toward the top. Note the closed air door over the secondaries. The throttle linkage and accelerator pump are on the left.

17

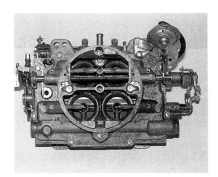
The top view of the AFB, production style. Primaries are toward the top.

The top view of the '72-and-newer Thermo-Quad. The primaries are toward the top.

There are several carb flanges used. This one has six holes rather than the more common four- or eight-hole flanges. Production carbs used these unique flanges but Edelbrock now uses an eight-hole flange on those carbs, which will fit any combination. With production hardware, be sure that your carb flange fits your intake manifold flange.

The top view of the Edelbrock AVS carburetor. The primaries are toward the top.

The housings look completely different once taken apart. This is an AFB production version with everything removed except for the velocity valve. To the left are the secondaries, primaries are to the right.

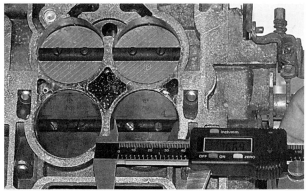
Carburetors are typically rated by airflow or not rated at all (production) so it is a good idea to check and record your actual throttle bore sizes.

sometimes 6). The Edelbrock aftermarket carbs have both patterns machined into the flange for more interchangeability (8 holes). The AFB and AVS carbs use a similar shape on the flange but the Thermo-Quad, and similar Rochester Quadrajet and Holley spread-bore carbs, use two very small primary bores and two very large secondary bores. The dual-pattern Edelbrock AFB-AVS carbs could be installed on Thermo-Quad manifolds but the large primaries on these carbs may interfere with the small bores used on these manifolds. There is an adapter that allows this swap to occur.

Throttle Bores—The round part of the carburetor that the air flows through is called the bore or barrel. On a one-barrel carburetor, there is only one throttle bore. On a four-barrel carb, there are four. The front two bores are called primary throttle bores and the rear two are called secondary throttle bores. At the bottom of the bore, there is a throttle blade or plate. This is used to regulate airflow. At idle, the throttle blade is almost horizontal, and at wide-open throttle, it is vertical.

The primary (front) throttles tend to be smaller than the secondaries (rear). They are both located in the bottom housing. All three carbs—AFB, AVS and TQ—have the throttle linkage located on the left side. The throttle blades or plates are designed to seal off the throttle bores when closed. They are mounted by two screws per plate to the throttle shaft. These parts are not serviced separately.

Most of the primary throttle bores are 1 7/16" and most of the secondary throttle bores are 1 11/16". The 750 cfm rated carbs have 1 11/16" primary throttle bores and the 800 cfm rated carbs use 1 3/4" throttle bores on both the primary and secondary. If you do not have a new Edelbrock carburetor, I would recommend measuring your

Carter/Edlebrock Components & Tech Specs

While some of the carbs have primary and secondary throttle bores that are about the same size, many have much smaller primaries than secondaries. Measure both.

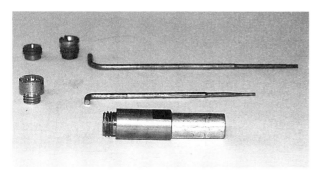

The three parts at the bottom are for the Thermo-Quad and the three at the top are for production-style AFB/AVS carbs. The Edelbrock uses the same jet in the primary and secondary locations, shown as the short jet on the extreme left. Note that the Thermo-Quad metering rod is much shorter than the AFB/AVS versions. The tall jet at the bottom is the TQ secondary jet.

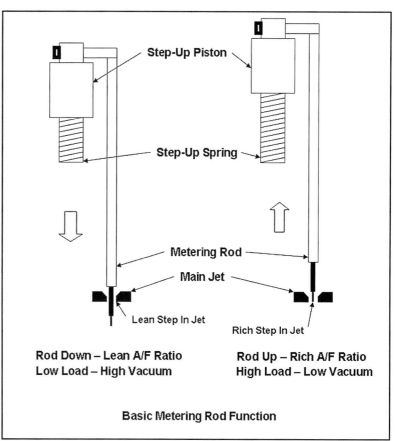

The basic metering rod function is shown above with the cruise mode shown on the left and the power mode shown on the right. The manifold vacuum pulls the metering rod down in the cruise mode. When the vacuum goes to zero in the power mode, the step-up spring pushes the piston up, which pulls the metering rod up and richens up the fuel curve.

throttle bores and writing it down in your notebook.

Metering Rods—The metering rod is unique to Carter/Edelbrock carburetors. The rod itself is small, straight and made of brass with a short 90° hook at the end. It is about the size of a toothpick, and installed vertically in the carb. The standard jet opening is somewhat larger than normal to compensate for the size/area of the metering rod. The area left between the outside diameter of the metering rod and the inside diameter of the jet is what determines how much fuel flows to the engine.

Additionally the metering rod has two or three diameters and is controlled by the engine's vacuum. At high engine vacuum, the metering rod is pulled down and the larger diameter has been pulled into the jet opening; this leans the ratio down for cruising and part-throttle operation. As the throttle is opened to wide-open throttle, the vacuum drops to zero and the metering rod rises, so that the small end sits in the jet and the ratio is richer, with more fuel being delivered.

In each carburetor there is a left and right metering rod, so they are usually serviced in pairs. Currently there are about 30 different metering rods available from Edelbrock. They range from 0.057" x 0.049" to 0.075" x 0.047". See the chart on pages 29 and 30.

The metering rod is really a team of four parts—the cover, the piston, the metering rod and the step-up spring. The pistons don't tend to wear out so they do not need to be serviced separately. There are two styles of AFB-AVS metering rods—the 2-step and the 3-step. The 3-step was only used in production. The 2-step was used in production and in all of the Edelbrock carburetors. The 3-step system can be swapped with the 2-step system if the

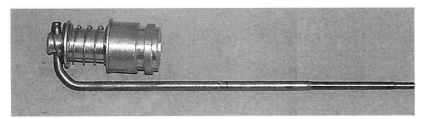

This is a 3-step metering rod but it is very difficult to see by eye. The two cruise steps are very similar in size and very hard to show in a photo.

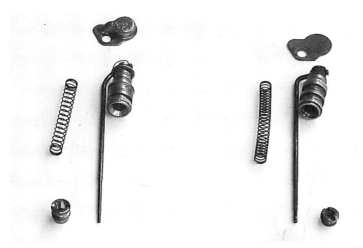

The 3-step metering rod hardware is to the left and the 2-step metering rod hardware is to the right. Note that the jet on the left is taller, and the cap at the top right is completely flat in the 2-step version.

This is the flat metering rod cover on the Edelbrock AFB/AVS.

This is the raised-center cap used on the production AFB/AVS carburetors with 3-step metering rods. The cap is in the center, held down by a single screw.

numbers were used on the rod itself. Since the metering rods can be changed without disassembling the carburetor, they make the best part to use to tune the engine package once the carb is assembled and installed on the engine.

Step-Up Springs—Step-up springs work with the metering rods and basically change when and how quickly the metering rod moves up or down which affectively changes the fuel curve in aspects of drivability rather than wide-open-throttle power. Edelbrock services five different springs, which are identified by color. The TQ used only one central spring but there is an additional adjustment on the central piston.

Covers—The cover goes over the top of the metering rod assembly and is held in with a small screw. The Edelbrock and some production versions are flat (2-step rods) and some production carbs have the center of the cap raised (3-step rods). The TQ uses a center piston and a long rod-arm that holds both metering rods, which makes it unique.

Metering Jets—Jets are used to meter the amount of fuel that passes through the carburetor passages into the throttle bore for any given amount of airflow. Typically there is one jet for each cylinder bore. Therefore in a four-barrel carburetor there are two primary jets or main jets and two secondary jets. The actual jet is quite small but is threaded and is removable. The jet is typically made of brass with a screwdriver slot across the top. The hole in the center allows fuel to pass through and is very carefully sized. There is a 45° chamfer on top. The jet size (diameter of the hole in the center) is etched in the top flange.

Main metering jets come in lots of sizes ranging from 0.077" to 0.119". They are serviced by Edelbrock. The AFB and Edelbrock AVS use the same jet style in both the primary and secondary sides. The 3-step metering rods used in some production applications uses a somewhat taller jet. They are interchangeable as long as you change the metering rods at the same time. The taller jets are no longer serviced. The Thermo-Quad uses unique jets, especially on the secondary side.

Jet Numbering—Each jet has a number etched into the top surface. The standard production AFB and AVS jets use a number 120 on the top half and 367 (0.067") for jets that have a size of 0.099 or smaller, or 409 (0.109") for jets that have a size of 0.100" and larger. With the Thermo-Quad, the primary jet numbers look like 120.4098 (0.098") or 120-4100 (0.100") while the secondary jets used numbers like 120-5149 (0.149"). If you have a jet that you can't read the number on, use the number drill set to measure the actual size.

metering rod cover (2-step is flat) and jet (3-step uses a taller jet) are changed. The 2-step hardware is readily available. The TQ metering rod system is unique and it is much shorter than the AFB/AVS style metering rod. It is no longer serviced. The production metering rods had identifying numbers etched into the rod itself. The TQ rods used numbers like 75-2002 or 75-1998. Only the last 4

Carter/Edlebrock Components & Tech Specs

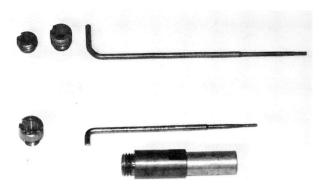

There are several styles of metering rods and jets. The parts on the top are for the AFB and AVS—the long metering rod. The parts on the bottom are for the Thermo-Quad. Note the much shorter metering rod. Also note the two jets shown in the upper left corner, that the one farther to the left is shorter than the one to its right. Also the TQ jet in the lower left is the primary jet and somewhat different in shape from the taller AFB jet. The TQ secondary jet is shown at the bottom.

On the top surface of the jet, there are etched/stamped numbers that indicate it size—in this case 120–368, or 0.068". These numbers can be hard to read so use a number drill to double-check.

These are the left and right primary clusters. While they look similar, they are not the same and must be kept in the correct locations.

Primary Venturi Cluster—For greater efficiency, a portion of the tube or throttle bore can be made with a smaller diameter, and that section is called a venturi. The air passing through the bore through this smaller area will speed up and create an area of lower pressure. This resulting lower pressure is used to suck fuel out of the bowls and into the throttle bore. Typically the venturi is 1/8" to 1/4" smaller in

These are the secondary clusters as used on the AFB and Edelbrock AVS. While they look similar they are not the same and must not be swapped from one side to the other.

Each of the secondary clusters has an air bleed located in the top surface of the cluster as shown by the arrows. The air bleed is located between the two attaching screws.

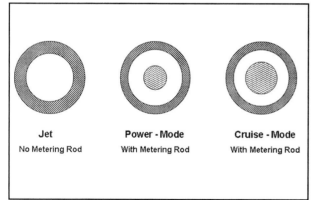

The effect of the metering rod on fuel flow. The jet on the left has no metering rod, like the secondary and the fuel flow is controlled only by the inside diameter of the jet. With the metering rod, there are two conditions. The power mode has no vacuum so the metering rod sits up, the small end of the rod sits in the jet. The difference between the two diameters is what controls the flow. In the cruise mode, the vacuum is high so the metering rod is pulled down and the larger diameter on the metering rod now sits in the jet and the resulting area is smaller than it is in the power mode and therefore less fuel flows and the engine is leaner for cruising, which is the desired condition.

diameter than the main throttle bore.

The primary cluster, one per primary throttle bore, sits in the primary throttle bores at the minimum cross-section or venturi. They are held in by two screws. They don't tend to wear out so they are not serviced separately. Each has a gasket that goes between the cluster and the main housing.

Secondary Venturi Cluster—Similar to the primary cluster, the secondary cluster sits in the secondary venturi. They are held in by two screws and are not serviced. The secondary cluster has an air bleed hole in the top surface. This hole should not be plugged. Each secondary cluster has a gasket similar to the primary. The secondary clusters are not used on the Thermo-Quad or the production AVS. On the Thermo-Quad and the production AVS, the secondary cluster is replaced by a nozzle bar.

Distribution Tabs—A distribution tab is attached to the side of the cluster—either primary or secondary. One of the tricks to production AFB 426 Hemi carbs is fuel distribution tabs. They stick off the side of the venturi in the throttle bores. In

REBUILD & POWERTUNE CARTER/EDELBROCK CARBURETORS

Distribution tabs are added to the sides of the inner nozzles on the primary and secondary bores on production carburetors. This is not done on Edelbrock carbs. It is done to improve the fuel distribution on a specific engine, carb and intake manifold package that has a cylinder-to-cylinder distribution variation.

The short main jet to the right is used in all 2-step applications, including all Edelbrock carbs, both primary and secondary, and the production AFB and AVS versions that used the 2-step metering rod setup. The taller jet on the left is only used in the primaries of the 3-step metering rod system.

The accelerator pump is used to spray extra fuel into the throttle bores to allow the engine to increase engine rpm. The nozzle sprays fuel into the top of the throttle bores (primaries only) and the pump is controlled by the accelerator linkage. The accelerator linkage has several adjustments in the upper link—generally 2 or 3 holes that move the link closer to the pivot. This is the AFB/AVS production pump. Accelerator pump link is to the left (arrow). Note the bowl vent to atmosphere on the end of the arm to the upper right (arrow).

The accelerator pump linkage on the Edelbrock AFB and AVS carbs. While it is very similar it doesn't have the bowl vent to atmosphere on the end of the extra arm (see above).

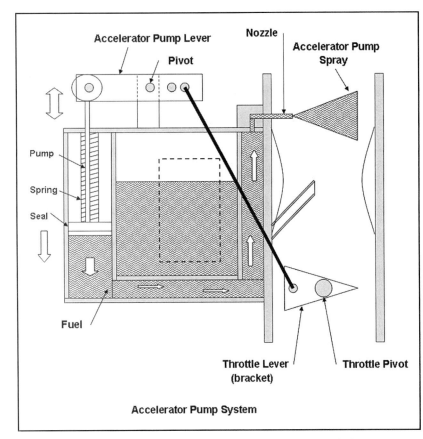

The accelerator pump is used to spray extra fuel into the throttle bores to allow the engine to increase engine rpm. The nozzle sprays fuel into the top of the throttle bores (primaries only) and the pump is controlled by the accelerator linkage. The accelerator linkage has several adjustments in the upper link—generally 2 or 3 holes that move the link closer to the pivot.

upper left photo above, the distribution tab is in the secondary (right side throttle bore) on the left and it points sideways at 10 o'clock. The size and location of these tabs is developed on the dyno. For performance purposes, these tabs can be removed. They are not used on Edelbrock carbs.

Note: Production engineers liked to add fuel distribution tabs to the carburetor, if there appeared to be a fuel distribution problem with the carb and intake package. In some cases, they also liked to use staggered jetting, where the jets on the left and right throttle bores are not the same. As more aluminum intake manifolds were used, Chrysler engineers used square-jetting and added the fuel distribution fixes to the manifold plenum. See Chapter 5 for more details.

Accelerator Pump—The accelerator pump is used to allow the engine to change speeds. At idle or part-throttle, the engine needs extra fuel to be available to be able to move up to a higher rpm. The

Carter/Edlebrock Components & Tech Specs

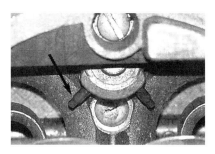

In the center is the accelerator pump nozzles on the AFB and AVS. The actual nozzle or shooters (one points to the left primary and one points to the right), are held in by one screw, just out of sight under the air horn lip. Edelbrock services several different sizes of nozzles so they can be tuned if required.

The nozzles on the Thermo-Quad are different from those on the AVS and AFB. Since the small primaries on the Thermo-Quad are farther apart than they are on the AFB/AVS, the nozzle angles are straighter while still pointing at the center of the primary throttle bores.

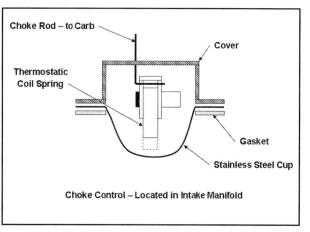

The typical production choke uses a thermostatically controlled spring and rod mechanism to activate the choke. The thermostatic device looks like a coil spring but is actually a coiled, bi-metal strip designed to change the height of the choke rod as the engine warms up or is cold. This style of choke requires a heat cross-over in the intake manifold to provide the basic engine heat input to the thermostatic device. It is typically located on the right side of the intake manifold.

accelerator pump provides this extra fuel. The pump squirters or nozzles are located above the clusters and are only used on the primaries. The accelerator pump itself is controlled by a linkage that attaches to the throttle linkage, typically on the left side of the carburetor.

Generally speaking, the actual accelerator pump, with seal, will come in your rebuild kit. There are several different arms and associated linkage. Edelbrock services three different accelerator pump assemblies based on the carburetor's part number. They have a separate one for the marine applications.

Accelerator Pump Nozzles—The AFB and AVS use the same basic accelerator pump nozzle and gasket. Edelbrock services these nozzles in 3 sizes—0.024", 0.033" and 0.043". The TQ uses a unique nozzle. The accelerator pump nozzle squirts fuel into the primary throttle bores. While you could drill them out to larger sizes, I would not recommend this modification because so many sizes are available from Edelbrock.

Bowl Vent—A bowl vent is used on the production AFB and AVS carbs. A bowl vent is not used on the Edelbrock versions. The standard bowl vent is part of the accelerator pump's linkage. It is not serviced separately. The TQ vents the fuel vapors to a fitting so it can be connected to a canister and recycled.

Choke—When the engine is cold, it requires a relatively rich fuel mixture in order to start. This condition is caused by the fuel (gasoline) not being able to vaporize before it reaches the cylinders until the intake manifold heats up. This means that the carburetor has to supply a much larger quantity of fuel when starting and running a cold engine. Additionally the velocity of the air through the throttle bores is low during starting which lowers the venturi effect. Cold enrichment is accomplished by applying the choke valve or plate, which is located above the primary venturi. When the choke plate is closed, the vacuum in the throttle bores is high, and this draws more fuel through the main nozzles and idle ports. There are manual and electric chokes used on these carburetors.

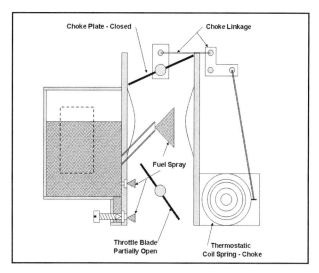

The choke operation only comes into play when the engine is cold. Cold, the choke plate across the top of the throttle bore is closed. As the engine warms up after starting, the bi-metal spring pulls the choke plate off so that it sits vertically and has no effect on the airflow through the throttle bore. In an electric choke, an electric device replaces the bi-metal thermostatic coil spring.

23

The bi-metal coil spring sits in a well area heated by the exhaust heat-crossover below the bolted down plate (arrow). The rod that is running straight up goes to the actual choke blade mechanism.

The typical electric choke on the lower right pulls the rod that actually opens/closes the choke blade on the carburetor.

The hollow float as used in the typical AFB and AVS. The pin and pivot is to the left.

Note: It might be more accurate to say that the manual choke is a thermo-mechanical design because it is controlled by a bi-metal thermostatically designed coil that operates the rod that opens and closes the choke blade. The electric choke might be described as an electro-mechanical device because the electric unit still operates the rod that opens and closes the choke blade.

There are manual chokes and electric chokes. They come with the carburetor and are not serviced separately. Production chokes would be classified as a manual choke but they are controlled by a bi-metal spring so they are considered thermostatically controlled. Full manual chokes can also be used.

Floats—The carburetor is designed to regulate the amount of fuel that is delivered to the engine. To do this the carburetor must have a reservoir that is maintained at a constant level. This reservoir is called the bowl or float bowl, and the level in this bowl is controlled by the float. The carburetor has a float-operated needle and seat located in the fuel inlet that controls the fuel flow into the bowl. When the fuel in the bowl drops below a preset level, the float lowers and this allows the needle valve to open and fuel enters the bowl. When the fuel gets to the preset height, the rising float forces the needle against the seat and shuts off the fuel flow.

There are two floats in each carb, one on each side. The AFB and AVS use the same float. The float is held in by a pin, which also serves as the pivot. Floats are serviced by Edelbrock. The TQ float is unique. It is black and made of a special material called nitrophyl.

Needles and Seats—Typically a new needle and seat will be part of your rebuild kit. The needle and seat control the flow of fuel into the carb or float

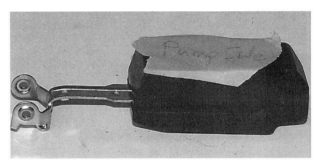

The Thermo-Quad float looks a lot different from the AFB version. It is made from a special material that is light but not hollow. It is always a good idea to label the floats left and right or pump side so they go back in the same location as original.

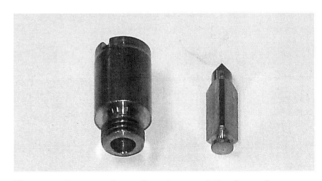

There are two needle and seat assemblies in each carburetor, one on each side. The needle is to the right and the seat is to the left.

bowl. They do wear out and get dirty. Edelbrock services both with a 0.0935" seat (standard) and a high-flow 0.110" seat.

Nozzle Bars—Nozzle bars are the fuel delivery method in the Thermo-Quad carb. They are not used in the AFB or the Edelbrock AVS. They are used on the production AVS. They replace the secondary clusters, but do not look anything like a cluster. They are located only over the secondary throttle bores.

CARTER/EDLEBROCK COMPONENTS & TECH SPECS

The nozzle bars are hard to see/find once the carb is assembled. With the secondary throttle open and looking up, you can see the nozzle bar going across the bottom of each throttle bore at an angle. They sit above the throttle blades and below the air door. They are used on the Thermo-Quad (shown) and on the production AVS.

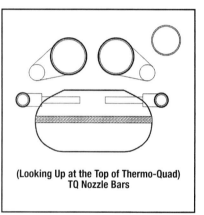

It is best to leave the nozzle bars installed into the housing. There is a left and right nozzle bar but they are only used in the secondaries.

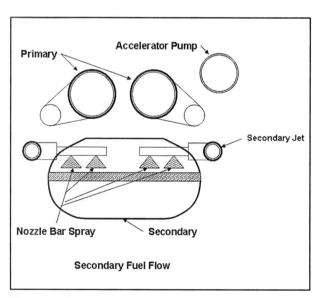

The nozzle bar has several holes in it (typically 3 or 4) and each hole sprays fuel into the secondary throttle bore.

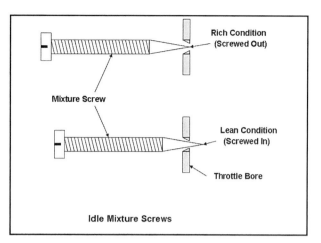

The idle mixture screws can be adjusted in and out to change the idle A/F ratio from rich to lean or vice versa.

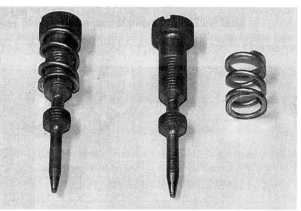

There are several styles of idle mixture screws. Most carburetors have two mixture screws—left and right—except for the production AVS, which has only one screw, and it is in the center and has a left-hand thread. The small spring going around the screw as shown on the left.

Idle Mixture Screws—There are two mixture screws on each carburetor, a left and right on the primary only. They sit in the center-front section of the carb toward the bottom or baseplate. They usually have a small coil spring around the outside of the screw. The heads are slotted for the standard screwdriver. They are used to adjust the engine's idle characteristics. The TQ's mixture screws are slightly different. On production carburetors, the idle mixture screws are often located under the plastic cap which should be removed before trying to set idle.

Carb Linkage—Carburetors have many linkages on them for the choke, the accelerator pump and the secondaries (not to be confused with the throttle linkage, page 73, top left). Edelbrock services the basic accelerator pump linkage and offers the special adapters for the Chrysler and Ford throttle linkages.

Secondary Linkage—The AFB, AVS and Thermo-Quad carburetors are all four-barrel designs, which means that there are two primary throttle bores (front) and two secondary throttle bores (rear). Typically the secondary linkage is staged so that the primaries open first and then the secondaries. For racing, it is sometimes changed so that they are 1:1 or they both open at the same

The basic carb linkage comes assembled with the carburetor but the accelerator pump linkage, center, generally has three adjusting holes so it can be fine-tuned. It is in the center hole as shown (arrow).

The Thermo-Quad bowl vent must have the link in the center of the Y. If you disassemble the carburetor, you must make sure that it is in the Y when you reassemble it.

The actual transfer slot in the primary (arrow). It is in the center, just above the tip of the throttle blade and just below the idle mixture screw—the small vertical slot.

This is the AFB velocity valve, which is located above the secondary throttle blades. This one is from a production carburetor—a small cfm model.

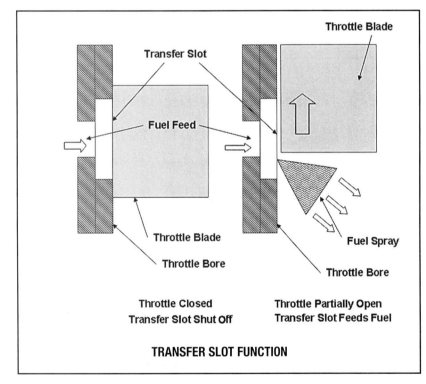
The transfer slot is very small and is not used on all models but it is designed to help drivability. The transfer slot is a slot that is cut vertically in the side of the throttle bore. It must be very carefully positioned because the throttle blade should cover it when the throttle blade is closed and it should start opening just as the throttle blade starts to open. As soon as it is uncovered, it sprays fuel to help to transition to a new engine speed.

The AFB has an inertia or weighted air valve but it isn't as easy to adjust.

Transfer Slot—A transfer slot may not be used in all carbs. It is more popular in production-based carbs. It is a small, vertical slot machined into the throttle bore just above the tip of the throttle blade—primary only in most cases. It is used to help drivability.

Velocity Valve—The AFB carburetors use a velocity valve above the secondary throttle blades. It is rolled open by the flow of air down the throttle bores. There are a couple different shapes if production versions are included.

Carb Gaskets—The carb gaskets that seal one housing to the other are included in the rebuild kit. There are slight differences between the production AFB and AVS and the Edelbrock versions. Edelbrock also services these gaskets separately. The TQ uses two carb gaskets (one is larger than the other) because of its three-piece design. The base gasket that goes between the carb and the intake manifold may also be included in the rebuild kit but they are also serviced separately. The TQ

time. This is not recommended for any street or dual-purpose application. The engine can not use this much airflow at low rpm. The AVS and TQ have a spring-adjusted air valve that allows for easy adjustment of the airflow through the secondaries.

Carter/Edlebrock Components & Tech Specs

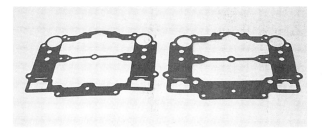

There are several versions of the AFB main body gasket that come in the typical auto parts store rebuild kit. Note that while the top part of the two gaskets is similar, the lower section is larger in the gasket on the right and also somewhat different in shape.

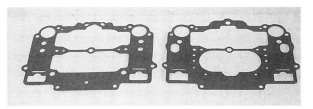

The main body gasket on the left is for an AFB while the one on the right is for the production AVS. Note that the upper sections are similar but the lower section on the AVS one, to the right, is somewhat oval while the AFB one is somewhat rectangular. Obviously the Edelbrock AFB and AVS units use the same gasket.

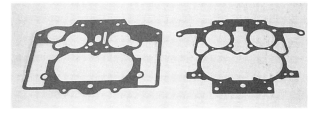

The '72-and-newer Thermo-Quad uses two main gaskets because it is a three-piece design.

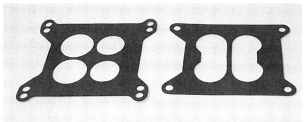

There are many styles of carburetor base gaskets—the one that goes between the carburetor itself and the intake manifold. The one on the left is a four-hole gasket and the one on the right is a two-hole design with the primaries to the right.

These could be considered spacers since they are about 0.200"–0.250" thick. The one to the left is made of thermoplastic and uses a gasket on top and bottom while the Thermo-Quad version on the right has the gasket built into the upper and lower surface.

gaskets are unique from the AFB/AVS units in shape—two small and two large throttle holes—and in thickness. The original TQs used a normal gasket and a thermoplastic spacer, about 0.200" thick. This seems to have been replaced by a one-piece assembly with the gasket printed on the top and bottom.

Note: Two gaskets and a plastic insulator is a better overall package than a metal spacer. The thermoplastic spacer, installed between the carb base and the manifold, helps keep heat away from the fuel which can be very helpful during the summer months and during cruising events.

Jetting Packages

Jetting packages used to be offered for specific racing combinations—engine, manifold, cam, headers, etc. The kits would have the jets, metering rods and accelerator pump parts needed to make max power and torque based on a the combination of power parts. These kits are extremely specific. Edelbrock now offers their calibration kits, which include several jet sizes (up to five sizes) and several metering rods (up to seven) and an assortment of step-up springs. That way one kit can work for many similar engines, packages and applications.

In the 1970s and early 80s, Carter used to sell "Strip Kits" for the Thermo-Quad and other models. These Strip Kits offered a great selection of jets and metering rods for endless tuning, but they are no longer available new. However, you might still find them at car swaps or online resources such as eBay.

Throttle Linkage

Once mounted onto the manifold with the engine installed in the car, the carburetor is typically connected to the vehicle by a throttle cable. Because of the accelerator pedal and bracket, this throttle cable is unique to each body style. Cables do need to be lubricated on occasion especially on older vehicles; they generally only need to be replaced if they have been damaged.

With the Carter/Edelbrock carburetors, usually the carburetor's throttle linkage will clear the manifold and any spacer that might be used. Sometimes a piece of the linkage interferes with the top of the manifold or the side of the plenum. It is always a good idea to check this out as early as

To use the universal Edelbrock carburetors on Chrysler engines, you have to use a linkage adapter as shown. They are available from Edelbrock. Ford engines also require an adapter.

The throttle cable comes from the car's firewall/accelerator pedal and is clamped in a bracket at the rear of the intake manifold. You always want to clamp on the metal end of the cable.

The side view of the throttle linkage shows the adapter, the cable and the transmission kickdown linkage all set up next to one another.

The throttle linkage can get pretty busy on the left side of the carburetor. The cable comes forward from the bracket to the center pivot. The throttle springs go forward from the pivot. On automatic transmission cars, the transmission kickdown linkage is the slotted rod on the right extending to the top.

possible. Since Edelbrock carburetors are built to fit all production engines, they offer special linkage adapters for Chrysler and Ford vehicles that allow the use of the standard carburetor.

The throttle cable is typically supported by a bracket that mounts to the rear of the manifold. In many cases, this throttle cable bracket is held on by the intake manifold attaching bolts. This requires the bracket to straddle across the top of two intake runners (typical of wedge-head configuration). This bracket can have clearance issues with today's bigger intake ports, and/or raised intake ports.

There are several sources for linkage kits for various models that are similar to production hardware. Because this hardware also doubles for restoration applications, you can find much of your parts at resto sources like Year One.

Special linkage applications like the eight-barrel systems, both inline and cross-ram, can also be found at the same resto suppliers as above as long as the eight-barrel system was used in production. If not, there are several sources for

The early-style dual four-barrel inline carburetor linkage allowed one carburetor to open somewhat ahead of the other and then get caught up at wide-open throttle. Allowing the one carb to open first is a better setup for street applications.

A newer style eight-barrel linkage is more solid and would seem to be more desirable for racing or manual transmission use.

race or universal linkage kits, such as Edelbrock, Mr. Gasket or Lokar Performance Products. There are several styles of eight-barrel, inline linkage systems available as well, depending on the application.

BASIC CARB SPECS

In the production carburetor group, there are hundreds of part numbers and each one has unique specifications. This can be very confusing. So for reference, I've selected a few typical production carburetors from each group—AFB, AVS and TQ. For these carbs, I have listed the typical main jet, secondary jet and metering rods. This spec sheet will provide a starting point or provide a baseline if you get lost in your tuning adjustments.

Tip: I would recommend saving this chart for future reference. It can be handy to come back to it for any number of reasons, especially when calibrating the carb in Chapter 7. If you get too far off track with your adjustments, you can always refer back to this chart to start over.

BASIC CARBURETOR CALIBRATION
(Production Carbs)

	Engine	Part Number	Main Jet	Secondary Jet	Metering Rods	Step-Up Spring
Carter AVS Series						
650 cfm	225–327	1801, 1802 *	0.089"	0.089"	0.065" x 0.052"	Orange
800 cfm	302–400	1805, 1806*	0.095"	0.098"	0.068" x 0.047"	Orange
Carter AFB Series						
750 cfm	225–327	1403, 1404 **	0.089"	0.0689"	0.065" x 0.052"	Orange
600 cfm	426 Hemi	1406 **	0.089"	0.080"	0.075" x 0.047"	Yellow
Carter TQ Series						
750 cfm	350–502	1407 ***	0.096"	0.143"	0.069" x 0.039"	NA
800 cfm	350–502	1412, 1413 ***	0.096"	0.143"	0.069" x 0.039"	NA

* Production AVS carbs are no longer available, so Edelbrock numbers were substituted for similar applications. The specs are based on production versions, not the Edelbrock numbers.
** Production AFB carbs are no longer available, so Edelbrock numbers were substituted for similar applications. The specs are based on production versions, not the Edelbrock numbers.
*** Production TQ carbs are no longer serviced, so these are approximate Edelbrock numbers for this application. The jets and metering rods reflect what was used in production, not what might be in the Edelbrock versions—see the next page for Edelbrock-specific information. The overall Edelbrock specifications summarized on the next page tend to be much more detailed than the specs available for production carbs.
Note: Many production AVS and Thermo-Quad carbs use a 3-step metering rod so these specs represent the largest and smallest steps.

EDELBROCK

This Edelbrock carb chart is based on the spec sheets and the basic engines for which they are recommended. This can provide a baseline or starting point for any performance project or a quick reference for the carburetor that you just purchased.

BASIC CARBURETOR CALIBRATION
(Edelbrock)

(AVS)	Engine	Edelbrock Part Number	Main Jet	Secondary Jet	Metering Rods	Step-Up Spring
Thunder Series						
500 cfm	225–327	1801, 1802	0.086"	0.095"	0.065" x 0.052"	Orange
650 cfm	302–400	1805, 1806	0.095"	0.098	0.068" x 0.047"	Orange
800 cfm	350–502	1812, 1813	0.113"	0.107"	0.068" x 0.047"	Orange
(AFB)						
Performer Series						
500 cfm	225–327	1403, 1404	0.086"	0.095"	0.065" x 0.052"	Orange
600 cfm	302–400+	1406	0.098"	0.095"	0.075" x 0.047"	Yellow
750 cfm	350–502	1407	0.113"	0.107"	0.071" x 0.047"	Orange
800 cfm	350–502	1412, 1413	0.113"	0.101"	0.071" x 0.047"	Orange

Chapter 3
Carb Selection

With any car project, you want your engine compartment to have the proper look and the carburetor(s) sitting right on top.

While selecting a carburetor could be one of the first items considered in an engine project, that is not always the case. Today you can order a crate engine and it comes already assembled, usually without an actual carburetor. However, many aspects of the engine, like displacement, intake manifold, overall engine output and usable engine rpm range, should be considered when selecting a carburetor. There are many other aspects that can come into play as well, like cost, availability, adjustability and the type of fuel that will be used.

Rebuild or Replace?

If you have a complete original engine that you plan to rebuild, then you may also have the original carburetor. This question may also be looked at as "new or used?" A used engine tends to have to be rebuilt because of wear to the rings, bearings, etc. but a carburetor may not require any hardware changes depending upon its condition. If in doubt, and it is an AFB or AVS, a new carburetor may be the answer. With a Thermo-Quad, you have to rebuild it or switch to an AVS or AFB. If you have a 1971 Thermo-Quad, use any of the '72-and-newer Thermo-Quads or replace it with an AVS.

Calculating Carb CFM to Engine Displacement

Although the combined engine package is critical to carburetor selection, the most important aspect is the engine's displacement, because this is directly related to the cfm that the engine flows and the rpm where it makes the

This is a '71 Thermo-Quad and it should be replaced because service parts are very hard to find in usable condition.

most power. Cylinder heads, cams, intake manifold and exhaust hardware are all important and are discussed beginning on page 34.

There are so many different sizes of four-barrel carburetors that you need to be somewhat more specific than just "four-barrel." One carburetor is not best for all applications. To this end, I have come up with the General Carb Selection Chart on page 33 that is based on the equation on page 32. The chart lists what size carburetor to use with which

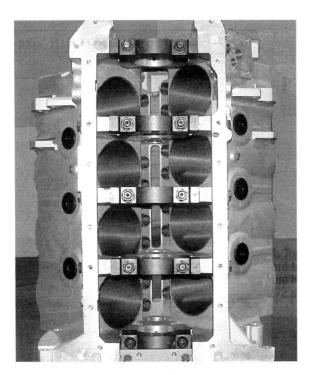

When carburetors were commonly used in production, engine displacements were usually similar to the production engines. Today with aftermarket parts such as this big-bore 440 Mopar block, the displacements can get much larger, and the carburetors need to flow more air and also flow more fuel for the added horsepower.

VOLUMETRIC EFFICIENCY

Volumetric efficiency is defined as the actual airflow that is measured at each rpm divided by the theoretical capacity (cfm) at the same rpm. To convert this V.E. number to a percent from a decimal, multiply by 100.

V.E. = actual cfm (measured) ÷ theoretical cfm (calculated) x 100

Note: The theoretical cfm is calculated by the carb cfm equation on this page.

Based on my experience, the following chart shows some approximate volumetric efficiency numbers.

Engine Package	Approximate V.E.
Stock:	60–80
HP Dual-Plane:	75–90
Single-Plane:	90–100
Race Dual-Plane:	95–105
Race Single-Plane:	105–115

displacement engine, for performance/racing applications. The basic equation is as follows:

carb cfm = (displacement x rpm ÷ 3456) x V.E.

where:
displacement is in cubic inches
rpm is the peak or shift speed
V.E. is volumetric efficiency, which is 1.0 (based on the chart above right)

Note 1: I list the engine rpm as peak or shift speed, but you can tune for any specific engine speed.

Note 2: I list the VE as 1.0, which is an average approximation of single-plane (90–100) and race dual-plane (95–105) manifold styles.

For example, if you have a 350 cubic-inch engine with a V.E. of 0.9 using a standard dual-plane intake manifold and a hydraulic cam, which has a peak horsepower rpm at 5500, the carb cfm would be calculated as:

(350 x 5500 rpm ÷ 3456) x 0.9 = 501

You would probably select a small 600 cfm carb.

What You Need to Consider

Engine Application—What you plan to do with the engine is a very important aspect of the carburetor selection process. While there may be many sides to your application, it tends to come

Race engines can easily, albeit expensively, be tuned to run rpm as high as 8000 rpm or more. These engines are not generally good for the street or dual-purpose applications, so I have not put that column on the chart. Small displacement race engines can also be tuned to run 6000–7000 rpm, but this high-rpm hardware is not typical of stock or production small displacement 4 cyl., 6 cyl., and V8 engines. Also, if the calculation indicates a carb cfm of 955 or 1000, do not automatically assume that you'll need a 1000 cfm (2 x 500) dual-inline system. An 800 or 850 cfm carb may still be the best choice for your application.

down to whether you are going to run it on the street, race track or both. For street or dual-purpose applications, you need more flexibility. So if your calculation or the chart on page 33 indicates a 700 cfm carb would be best, you might be better off with a 600 cfm carb. For racing, cars with automatic transmissions tend to work like street cars (so select on the conservative side). Only race cars equipped with a manual transmission can take full advantage of the extra cfm.

Fuel—As discussed in Chapter 1, the type of fuel you are going to run is very important to carb selection, because either of the two types of alcohol or E85 would require very large jets, and this might mean a specially prepped carb, which Edelbrock does

Carb Selection

The WCFB four-barrel carburetor was used on the early 331-354-392 Hemi engines, now called Gen I. This was a dual-inline eight-barrel system from the Chrysler letter series cars of late 1950s.

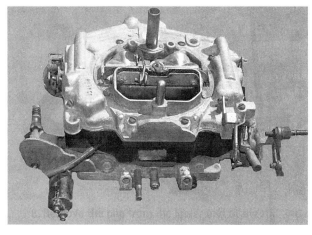

This is the '71 Thermo-Quad. You can identify it by the large single vertical pipe in the center-rear of the carburetor. Parts in usable condition are very hard to find. It should be replaced with a newer Thermo-Quad or by the AVS, using the Thermo-Quad adapter.

GENERAL CARB SELECTION CHART

Engine	Displacement	Carb cfm@rpm			
		4000	5000	6000	7000
4 cyl.	122 (2.0L)	141	176	(1)	(2)
	152 (2.5L)	176	220	265	(2)
6 cyl.	200	229	286	(1)	(2)
	225–230	260	326	390	(2)
	250	289	361	433	(2)
V8	270–285 (5)	316	395	474	(2)
	300–330 (5)	364	460	552	(2)
	340–360 (5)	(3)	506	607	709
	370–400 (5)	(3)	557	668	780
	425–455 (5)	(3)	637	764	891
	500	(3)	723	868	(4)
	550	(3)	796	955	(4)

(1) Small 4 and 6 engines (two-valve) stock hardware generally not ready for 6000 rpm

(2) Small engines (two-valve) generally not set up for 7000 rpm

(3) Large high-performance engine generally not set up for 4000 rpm

(4) Long stroke, large displacement engine not set up for 7000 (hi-rpm)

(5) On these V8s that list a range of displacements, I used the middle value or average to calculate the cfm.

Note 1: Carb selection for large displacement—426 to 500 cid—engines calculates to about 800 cfm. Larger 528, 572 and 600+ cid engines are somewhat easy to build today. These would calculate to 1000 cfm. This would indicate two carbs like two 500 cfm units—2 x 500 = 1000 cfm. Two 600s would be 1200 cfm and two 800s would be 1600 cfm. Edelbrock recommends two 500 cfm carbs for a dual-inline system. The 426 Street Hemi (1966–'71) used two 600 cfm units (approximately). Remember that in carburetors, bigger is not always better.

Note 2: Any carb cfm calculation for rpm or specific displacement not shown in the above chart can be easily found by using the equation listed on page 32.

Note 3: For any calculated carb cfm below 500, I would recommend selecting a 500 cfm AVS carburetor because the spring-adjustable air door used on the AVS allows you to tune for smaller air flows required by small displacements.

offer. The other aspect to fuel choice is the type of gasoline that is going to be used. The answer to this question is usually directly related to the engine's compression ratio, which dictates the fuel's octane capacity. This aspect is discussed in page 36 and Chapter 6 and can affect the calibration and jet sizes.

V8 Two-Barrels—Almost all V8 engines that used a two-barrel carb in production also offered a four-barrel package for that same engine so upgrading to the four-barrel package is easy. Additionally Edelbrock offers aluminum four-barrel intake manifolds for most of the popular V8 engines.

V6 Engines—Two-barrel carbs were common on production V6 engines. For many of these engines, Edelbrock offers an aluminum four-barrel replacement intake manifold that would allow this conversion.

I-6 Engines—Inline six-cylinder (I-6) engines offer a bigger challenge for the four-barrel conversion since the vast majority of these engines were built with one- and two-barrel carburetors. However, Clifford Performance offers an aluminum four-barrel intake manifold for the Chrysler Slant Six, the Chevy/GMC I-6, the Jeep 258 and the Ford I-6.

REBUILD & POWERTUNE CARTER/EDELBROCK CARBURETORS

The typical dual-plane intake manifold has two levels—an upper and a lower. Note that the two middle runners in the lower manifold face drop below the two end runners as they swing toward the center. Four of the cylinders tap into one level and the other four go to the other level.

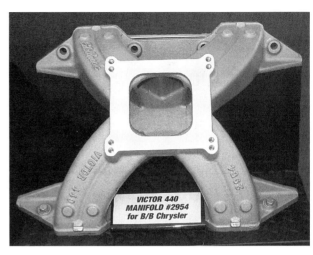

This is a typical single-plane where all runners meet in the center (plenum) on the same level.

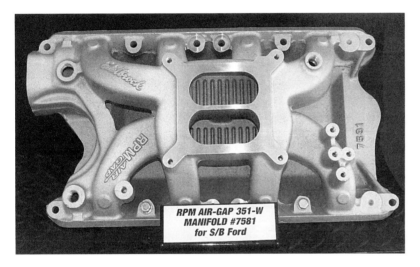

This is also a dual-plane manifold that joins the two center runners on the lower flange and the two end runners on the upper flange. The unique aspect is that this version has a cast plate located below the runners that keeps hot oil in the tappet valley from hitting the bottom of the runners and allows air to flow under the runners. These are called Air Gap manifolds by Edelbrock. Note the triangle to the left of center, below the word Edelbrock and the triangle to the rear to the left of the numbers 7581. Air can also get in on the sides between the individual cylinder runners.

WCFB—The WCFB carbs are very old and should only be used for resto applications. I will not discuss them in detail in this book. While the obvious upgrade for any WCFB package might be to the AFB, I would recommend an AVS.

'71 TQ—The '71 Thermo-Quad is superseded by the '72 TQs. Parts and tuning details for the '71 are very difficult to obtain. Because of the solid-fuel aspect of the newer TQs, you cannot copy the tuning tips for the '72 and newer carbs into this '71 unit. Because of the lack of parts, I will not discuss the '71 TQ versions in detail and will focus on the '72 and newer TQ's.

Intake Manifolds—There are two main styles of single four-barrel intake manifolds—the dual-plane and the single-plane. Most production engines used the dual-plane, which has two levels—one on top of the other—and four cylinders drawing off each level. The dual-plane design makes for a more complicated casting but they make more torque and offer better drivability. The single-plane manifold has all eight cylinders drawing from the same level or plenum. The single-plane tends to make more power but less torque and drivability is not as good. With a single-plane design, the size of the plenum, where the eight runners come together, can be an issue. The carburetor has to work with the intake manifold as a team, so the manifold's basic performance characteristics are very important. Another aspect of the manifold is the carb pad. There tend to be two—the standard four-barrel and the spread-bore (Thermo-Quad). Most AFB and AVS carbs fit on the standard four-barrel carb pad and the Edelbrock version offers both standard patterns machined into the carb's flange. The typical spread-bore or Thermo-Quad manifold had special casting because of the very large secondaries. If you have one of these manifolds, you can use the Thermo-Quad adapter (Edelbrock) to join the two parts.

Race—As long as the parts are available, you can race virtually any carburetor. Edelbrock makes the carburetors and race-required parts (jets and metering rods to richen carbs for increased power) are readily available. The trick tends to come in specific race classes. Since most of the engines are

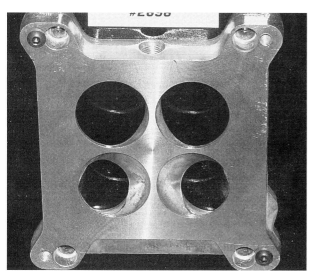

To adapt a Thermo-Quad to any standard carburetor manifold, you need this adapter. It can be used in the other direction if you want to bolt an AFB/AVS carburetor to a Thermo-Quad manifold.

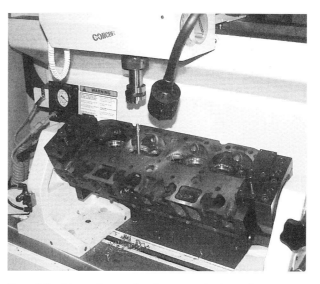

One of the biggest aspects of your engine's performance is controlled by the condition of the seats and guides in the heads. Having the guides redone and a fresh valve job done on the seats and valves can gain a tremendous amount of performance. The carburetor is often blamed for poor performance, often when it has nothing to do with the real cause.

production based, NHRA and IHRA Stock and Super Stock classes require that the stock production carburetor be used. In some cases, like the 426 Street Hemi ('66 through '71) the original carbs have been superseded by the similar Edelbrock version. In some crate motor drag-race classes (mainly IHRA), there are specific carbs listed as being approved for racing. In some cases, a carburetor replacement isn't approved because it hasn't been submitted. In all cases, contact your sanctioning body first to see what carburetors are legal and see if any superseded versions have been allowed.

Condition of the Engine—The basic condition of the engine is often overlooked in your performance plan but it is very important to the overall success. This would tend to indicate that the engine is running currently and has some time/mileage on it, i.e. it is not new. This means that the condition of the valve seats, valve guides, piston rings, pistons, valves and gaskets is important. For example if you do a valve job on the valves/heads, then the engine will make more power and get better economy. If the rings are worn/broken or the valve guides or valve seals are worn, then the engine will burn oil and this hurts performance and can't be fixed by adjusting the carb. New pistons rings are available from Speed Pro, Hastings Piston Rings and Total Seal Piston Rings. And don't over look the gaskets especially the head gaskets and intake gaskets because they can directly effect blow-by and oil and air leaks. Gaskets are available from Fel-Pro and Cometric Gasket and your local speed shop or auto parts store.

Overall Engine Package—In order to select the best carb for your engine, you really need to consider the entire package. Because the whole engine pumps air as a team, the engine should be looked at as a team. All of the engine's components contribute to the overall performance. The carburetor sits on top of the whole package, not just the individual components, but there are so many different component possibilities that the total number of combinations is just about infinite. You don't want to use a stock cast-iron intake manifold and cast-iron exhaust manifolds with a big mechanical roller cam or use the stock production cam with ported heads and a race single-plane intake manifold. I have put together some general engine packages in the chart on page 36 based on the assumption that you install a better cylinder head, a bigger cam, higher flow intake manifold, headers and upgraded ignition system. These are just general guidelines. The chart is discussed in greater detail in the following sections, so refer to the chart as you read them.

Horsepower—In the chart on page 36, the engine's horsepower output is listed as hp/cid or horsepower per cubic inch. In the first column, high-performance production, the hp/cid is shown as 0.8, so if you have a 450 cubic-inch engine, then you would make 360 hp. This is probably 50 to 100 hp better than an actual production engine, even though it has very mild specifications.

POWER PACKAGE COMBOS

Component	HP/Stock	Dual Purpose	Brackets	Serious Brackets	Drag Race	
hp/cid	0.80	0.9 to 1.1	0.85 to 1.05	1.45	1.8	
Cylinder Head(1)	CI	CI/Alum.	CI/Alum	CI/Alum.	CI/Alum.	
Compression ratio	8.5	9.0/10.0	9.0/10.0	10.5/11.5	11.5/12.5	
Cam	Hyd.	Hyd./Hyd.	Hyd./Hyd.	Mech/Mech	Roller/Roller	
Duration (@ 0.050")	210	225/225	235/235	250/250	260/260+	
Valve Lift	0.425"	0.475"/0.475"	0.500"0.500"	0.600"/0.600"	0.650"/0.650"+	
Peak rpm	5500	6000/6000	6000/6000	7500/7500	8000/8000	
Intake Manifold (2)	DP	HR/DPHR/DP	SP/SP	R-SP	R-SP/R-SP	R-SP
Exhaust (3)	HP Manifolds	S-headers	Race Headers	BT Race Headers	BT Race Headers	
Ignition (4)	Electronic	Electronic	Electronic	Race Electronic	Race Electronic	
Vol. Eff.(%)	75–90	95–105	90–100	105–115	110–122	
Peak HP (rpm)	5200	6000	6000	7000	500+	

(1) CI = cast iron, alum = aluminum
(2) DP = dual-plane, HR-DP = high-rise, dual-plane, SP = single-plane, R-SP = race single-plane
(3) HP Manifolds = high-performance cast iron manifolds, S-headers = street headers, Race Headers = 4-into-1, BT Race Headers = big tube race headers
(4) Electronic—standard high-performance ignition, race electronic = MSD 7 or equivalent.

Actual horsepower output is related to displacement, and most of the equations give you a horsepower-per-cubic-inch number so the actual displacement is very important. In the last few years, both cast and forged cranks in long strokes have become readily available for most of the popular engines which means the carburetor has to flow more air and fuel.

Cylinder heads are very important to the engine's power output and there are many heads, both cast iron and aluminum, that are available for use on the popular engines that generally offer upgrades over the stock heads (increased airflow).

Cylinder Heads—The ports in the cylinder head are one of the keys to how much air the engine flows and how much power it makes and what cam should be used. Again, referencing the chart above, CI refers to cast iron and Alum refers to aluminum materials. The first column, HP Production only shows a cast iron head because the typical stock muscle car V8 engine did not use aluminum heads. There are generally several upgrades for cylinder heads in both cast iron and aluminum. The drag-race-only heads are probably ported, perhaps CNC-ported which may be true of the serious bracket heads also. Remember that better heads will flow more air and therefore make more power and the extra horsepower will require more fuel from the carb and perhaps more airflow as well.

Most of the numbers used to discuss cylinder heads come from tests run on a flow bench. As discussed earlier, today almost all flow benches use a test pressure of 28 psi, which helps keep comparisons more equal. There are intake flows in cfm and exhaust flows in cfm, and mid-lift flows etc. Actual cfm gains can come from modifications to existing heads or from purchasing new cylinder head castings that offer increased flow numbers. You would like to increase the port flow more than you increase the port volume but this is not easily done.

Compression Ratio—While compression ratio is a function of the cylinder head and block, the

Domed pistons as shown increase the compression ratio, generally over 11-to-1.

The dual-plane intake manifold tends to make more torque, which helps drivability and makes the carburetor's job easier.

The single-plane intake tends to make more power and use more engine speed and this can cause regular adjustments in the carburetors calibration.

piston is the main factor if you want to change the ratio by 1.0 or more. The '72-and-newer engines, had either 8.0 or 8.5 compression ratio. The aluminum heads are always shown with 1.0 to 1.5 ratio more than the cast-iron heads, because aluminum conducts the heat away from the combustion chamber, which keeps the engine output numbers basically the same. In other words, if your high-performance cast iron and aluminum heads both flow 280 cfm, then the extra ratio in the aluminum head package allows it to match the output of the cast iron head.

If you want to change your engine's actual compression ratio by any large amount, like one full ratio or more, than you will most likely end up changing pistons whether you want to go up in ratio or down in ratio. Domed pistons generally mean hi-ratio while dished pistons mean low ratio with flat pistons generally being in the 9- or 10-to-1 area.

Cam—Basically the next three rows in the chart all relate to the camshaft. The cam starts as a hydraulic design and then moves up to a mechanical cam package and then a mechanical roller. The duration and valve lift follow the rule that as the ports get better, the cam gets bigger, in both lift and duration. Remember that the cam must work with the cylinder head so if you plan on using a 0.500" valve lift cam, then be sure that your head flows well at 0.500", not at 0.800", and also compare the flow numbers at half valve lift—or 0.250" in this case. Cams are available from manufacturers such as Comp Cams, Crane Cams or Bullet Cams.

Peak rpm—This is an estimate at best. Most hydraulic cams will peak in the 6000 rpm area and the production-based package doesn't generally have the good hardware but there are exceptions. A mechanical cam package, which usually includes rocker arms and pushrods, could pick up 500 rpm over the hydraulic or as high as 1500 rpm. However with better parts, better heads, bigger cam, I picked the higher number.

Intake Manifold—The intake manifold recommendations in the chart are obviously based on a four-barrel and the typical high-performance design in production is a dual-plane manifold (DP). The next step up is the high-rise dual-plane (HR-DP) which are sometimes called race dual-planes. They are made of aluminum and are only available from the aftermarket, like Edelbrock. The next step up should be the single-plane (SP), but single-planes generally increase higher rpm power while hurting low- and mid-range torque. Therefore, there are good ones and not so good ones. Many single-planes are not as good as the latest high-rise dual-planes, but some are better, so you should research which intake is best for your specific engine.

Exhaust—Most V8 engines have one version that has good, high-performance exhaust manifolds. This is considered the baseline. The next upgrade is street headers (S-headers) which tend to have small 1 1/2" or 1 5/8" primaries. The next step up is called race headers which usually means primary tube sizes of 1 3/4" to 1 7/8". The next upgrade is listed as BT Race Headers (big-tube race headers) and a big tube might mean 2" and larger.

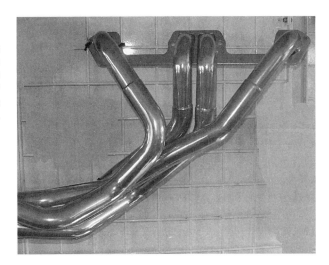

Smooth 4-into-1 headers will make more power but also require the carburetor calibration to be richened up.

Underhood air is warmer than outside air. Production air cleaners could have no snorkel, a single (shown here) or dual snorkels. Open air cleaners are much noisier than enclosed units similar to this one.

The 426 Street Hemi (1966 through 1971) used two AFB carburetors mounted inline. It was the last production use of the AFB. It is still popular in drag racing. This engine is a NHRA Stock class car.

Note: Race headers are generally 4-into-1 designs that have a smooth shape and big-radius bends for higher airflow. Many race cars today are using Tri-Y headers, but these designs are not commercially available at this writing (custom-made only).

Ignition—Like the other engine hardware, the ignition should be upgraded as the rest of the package is pushed for more output. However, the electronic ignition should be considered the baseline package even if it wasn't stock/production. While there are several options for race electronic, you could consider this to be an MSD7 or one of the digital ignitions from Accel or Crane.

Volumetric Efficiency (VE)—This is actually listed in the chart for reference only. It is somewhat of a summary of the parts listed in the column above it but it is also is similar representation to the last row on hp/cid (see VE sidebar on page 32).

Peak HP—The rpm that the engine package will peak is an estimate until you dyno-test it or run it in the car. However, to set up many of the systems in the vehicle, you need to utilize some approximations, so this gives you a starting point.

Racing or Special Engine Package Recommendations

Since any carburetor can actually be raced, this section is for the multiple-carburetor packages. I think that the inline packages were first into production in the mid-1950s. Next came the long rams, with the carbs outside the valve covers in the late-1950s and early '60s. Next came the true cross-rams on the '62–'64 413 to 426 cid Max Wedge Mopar packages. In 1964, '65 and '68, the 426 Hemi version of the cross ram was produced. Both are still being raced in NHRA/IHRA Stock and Super Stock today. The key to an eight-barrel package is the availability of an intake manifold.

Note: NHRA/IHRA Stock and Super Stock racing classes require that the production carburetor(s) be used unless they have been superseded and approved. The Edelbrock AFB carbs have been approved for the 426 Street Hemi. Carbs must be submitted to be approved, so this situation is constantly changing—check with your sanctioning body for latest details.

426 Hemi 8-Barrel Inline—Perhaps the best known today and the package that survived the longest in production (1966 thru 1971) was the 426 Hemi. It used two AFB mounted inline. It was rated at 425hp in production. This high performance street package was based on two approximately 600 cfm carbs. Production manufacturers did not like to rate the carburetor systems while Edelbrock uses the cfm-rating method almost exclusively. Several inline manifolds are available from Mopar for the 426 Hemi.

Chevy—There are eight-barrel inline manifolds available for both the small-block and big-block

CARB SELECTION

Edelbrock makes dual-inline manifolds for both the small-block and big-block Chevy engines.

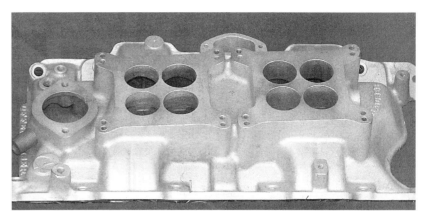

This is a dual-inline intake manifold for the Mopar small-block. Note that it is a single-plane style.

Dual-inline carbs are also popular on Ford engines. Manifolds are available for both the Ford small-block and the Ford big-block.

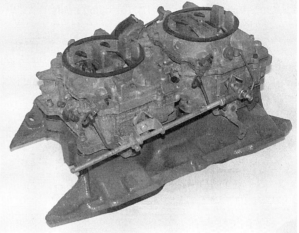

This is the production-based, dual-inline setup for the early Mopar wedge engines. It is also a single-plane. They were used on the early 1960s 383s and 413s.

Chevy engines. More recently Edelbrock has introduced an eight-barrel inline manifold for the early 348–409 Chevies—(actually fits the 348-409-427 W-series engines). These engines used an eight-barrel inline manifold in '60–'62. The new version is a dual-plane, high-rise design.

Note: There were several Chevy 265–283 small-block 8-barrel inline systems used in '55–'61, but they are listed as WCFB carbs. There is an 8-barrel inline manifold for the Chevy small-block available from Edelbrock.

Pontiac—The Pontiac 389–421 family of engines also used an 8-barrel inline package in the early '60s and Edelbrock now makes a new aluminum casting.

Ford—Many of the early small-block Ford performance packages used the dual four-barrel inline carburetion package. The '63–'64 Shelby 289 inline 8-barrel was perhaps the most common. Edelbrock makes a 289–302 intake and another for the 351 family.

Mopar—In the early 1960s up through about 1963–'64, there were several 383–413 big-block Mopar engines that used the dual 4-barrel inline system in production. These manifolds were single-plane cast iron designs—not what you'd like to use today for street or race applications. Perhaps the best known is the 343hp, 383 from 1962 and its bigger brother the 385hp, 413. Currently Edelbrock is making an aluminum dual-plane, high-rise dual-inline manifold for the 413-440 family of big blocks. Note: the 440-style manifold doesn't fit the 383—its too wide.

Mopar Cross-Rams—The '62–'64 Max Wedge intakes are still available. There is one made by Mopar and another resto part made by A and A Transmission. There is also a resto cross-ram made by A and A for the 426 Hemi. The Super Stock racers all use fabricated designs which are constantly changing.

Note: The 413 and 426 Max Wedge engines used very large-port cylinder heads which the manifold was designed to mate up with. Those big-port heads (cast iron) are being remade by Mopar today so the whole package can be swapped onto the much

39

The cross-ram intake system uses two AFB carburetor mounts somewhat across from each other. Note that throttle linkage uses a bell-crank mounted in the middle of the two carburetors. The foam around the velocity stacks is designed to seal to the hood scoop.

In the late 1950s and early '60s, Chrysler products used what was called the long ram system as shown here. These intake systems were designed for the wedge engines—383's and 413's. You can see that the carburetors actually sit outside of the engine's valve cover.

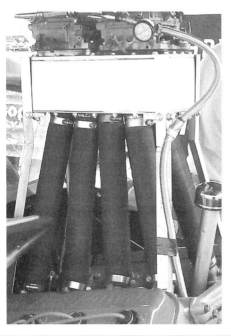

The first tunnel ram intake manifold from the '59 High and Mighty used Carter carbs. Tunnel rams became more popular with the 1970 Pro Stock drag racing class, but they aren't very popular today.

more readily available 440 short block. The '63 426 Max Wedge used the smaller 3447 AFB while the '64 426 Max Wedge used the larger 3705 AFB.

Mopar Long Rams—These late '50s and early '60s parts were designed to tune for a very low engine rpm so they are not compatible with today's engine hardware. They are best left to the resto projects.

Tunnel Rams—A tunnel ram intake manifold is an eight-barrel inline manifold but is very tall. The intake runners basically go straight up from the heads. While they make a lot of horsepower, they are not very popular unless the class is an all-out, max-rpm set-up like Pro Stock.

Note: Carter carbs were used on an early '70s NHRA Pro Stock package.

Recommendations Summary

Similar to a 1 3/4" throttle bore, the airflow rating of the carburetor is just a number. While it can be helpful to select the carburetor for various engine packages based on this cfm rating, it is not the only item that should be considered.

Cost—Before buying a new carburetor, check into the cost of the product. Carburetors of varying cfm sizes may not all have the same cost. For example, a 750-cfm version may cost less than an 800-cfm version.

Availability—There are two parts to the availability question—new parts (carburetors) and service parts (jets, metering rods). Since Edelbrock went into manufacturing these carbs, the parts are readily available for AFB and AVS

Adjustability—One of the aspects of any carburetion system is that you can rarely guess the full fuel-curve correctly at all aspects of the engine performance. Having a carburetor that you can adjust without taking the carburetor itself apart is a giant advantage. Additionally, having an easily adjustable spring-loaded secondary air door like the AVS (and Thermo-Quad) is another big advantage.

User-Friendly—If you have to send your carburetor to a professional or take your car to a professional tune-up shop, it is going to cost you a lot. So you would prefer to have a carburetor that you can adjust yourself. To accomplish this, you need an adjustment system that is user-friendly. The metering rods and air door adjustments are pretty easy to do and pretty easy to understand.

Chapter 4
Carb Rebuilding

The AFB comes apart easily; remember that there is a float on each side attached to the top.

Note: In this chapter I will cover the diassembly, rebuilding and assembly of the AVS, AFB and Thermo-Quad carburetors. In order to make it as easy as possible to follow, I've covered the step-by-step for each model with the text first, followed by its respective photo how-to. Read the text first, then move on to the photos.

Rebuild Kits & Parts

Before you try to take anything apart, you should get familiar with the parts—what they generally look like and how to associate the various terms with the specific hardware. Perhaps the most important aspect of any carburetor project is to be sure that you have a carburetor rebuild kit in hand BEFORE you start any disassembly. That way you know that you have the gaskets available to put it back together.

At this writing, rebuild kits are available for all three groups of carburetors—AFB, AVS and TQ. They are available at any auto parts store or performance parts dealer. They like to look up their kit numbers by using the number on the carburetor so go armed with several common numbers listed in the charts in Chapter 1. That way you'll have a better chance of finding the proper kit. They tend to service these kits in a one-kit-for-all approach so you end up with a few extra pieces, but one kit one number is much easier to stock than twenty. Remember that these kits generally do not come with any jets or metering rods. Edelbrock also has an AFB/AVS rebuild kit.

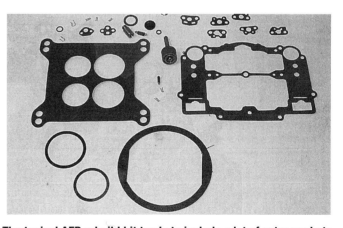

The typical AFB rebuild kit tends to include a lot of extra gaskets. One kit is used to service many numbers and versions so they include several variations and styles of gaskets. That means that you should keep the old gaskets until you rebuild the carburetor and use the old gasket to select your new ones.

Other Parts Needed—Assuming that you have used the information in Chapter 3 and have already chosen the carb best for your engine package and application, and have obtained the correct rebuild kit from Edelbrock or an auto parts store, you'll need to gather all the other parts needed. This would be any tune-up parts that you plan on using, like bigger (or smaller) jets, and metering rods. Edelbrock offers calibration kits that offer these parts in one package. If you plan on using a carburetor spacer, you should have it and the extra gasket(s) on hand. Other service items like carburetor linkage and fuel lines

The small parts of a carburetor are very small and very easy to lose or misplace. Place each part in a separate plastic sandwich bag and then label. Place the clips in with the links to help keep things straight. Also keep the old gaskets on hand until you rebuild so you can use them to identify/select the new gasket.

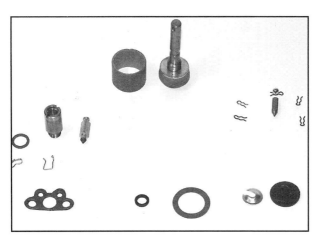

The key tool in identifying jets and nozzles is a complete set of drills, especially #1 through #60 number drills. These small-increment drill sizes allow you to figure out what size the various jets are because you can't always read the numbers on the jets themselves.

and filters and air cleaners should all be assembled.

Tip: Don't throw the old gaskets away because they can be very useful in determining the correct gasket to use for reassembly. This is especially true of small gaskets.

Shop Setup

While the carb is not very large, it is comprised of many small parts that can easily become lost on a cluttered workbench. Set the carburetor up on a repair stand or two pieces of 2 x 4. The height above the workbench helps protect the throttle plates and linkage from damage. Set the carb gaskets off to the side, out of harm's way but within easy reach. The AFB and AVS carbs are two-piece housings and both parts are aluminum castings.

Tools—The tools required for disassembling and rebuilding a carburetor are pretty basic—common wrenches, standard screwdriver assortment and standard and needle-nose pliers. Be sure to have a set of Torx drivers or bits if you have an Edelbrock version. A 1/4"-drive, small socket set can be helpful on a Thermo-Quad. The very important measuring tools are dial-caliper, steel scale, 0–1" micrometer, a complete set of drills, especially numbered drills (#1 through #60). I'd suggest having a box of sandwich-sized plastic bags available because there are a lot of small, tiny pieces on carburetors and these make handy places to keep all the related parts together. I would strongly recommend a pencil and notebook to keep track of everything, which is very important in tuning carburetors.

Make Your Own Data Sheets

General spec sheets for the carbs are listed in Chapter 2 on pages 29 and 30. The part numbers are listed in Chapter 1 beginning on page 11, and these can be helpful in getting the correct rebuild kit from an auto parts store. The production spec sheets are only very general and should be checked in detail against your specific carburetor. The Edelbrock specs can be matched by Edelbrock part number. In either case, you should make your own specification sheet for your specific carburetor. This sheet should contain any measurement or spec that is not already known as you disassemble it. This includes the primary and secondary throttle bore diameters and the venturi sizes, measuring the metering rods (there tend to be numbers on them but they can be hard to read) and jet metering passage size if you can't read the jet numbers on the top surface or want to confirm the size. Write everything down on the spec sheet.

Although tuning is covered in Chapter 6, there are certain things that you have to do before you put the carburetor back together that will affect future adjustments, calibration and tuning. Changing jet sizes, float height or velocity valve modifications are difficult to do after the carburetor is reassembled. While a spec sheet is a list of what you have, the tune-up is where you want to go, so I recommend you compile a second tune-up sheet that details the initial tuning parts installed to serve as a baseline to refer to later on. Once the engine is up and running, you might want to revert back to where you started in case you get "lost" while changing jets, rods and such. While the actual adjustments, or calibration, will be discussed in Chapters 6 and 7, I'll discuss some initial tuning steps beginning on page 117.

General Disassembly Notes

In this chapter, I am just going to discuss the disassembly and reassembly of these carburetors. As I mentioned, the adjustments to the carburetor are

CHANGING JETS

In Chapter 1, I discussed the A/F ratio, which is the key to the lean/rich mixture. If you want to lean-out the carburetor or richen it up, it is best to do it while you have it apart. Edelbrock offers some general guidelines for their basic carburetors, but there are so many engine packages and those will only get you in the ballpark. The trick is to estimate the fuel you need so you don't have to disassemble the carb again. The first step is to calculate the various jets' areas:

$A = 3.1417 \times D \times D \div 4$

Where A = area of the jet, D = diameter of the jet. The chart below lists some of the common answers for readily available jets and metering rods. For example, with a main jet diameter of 0.098, the area is:

$3.1417 \times 0.098 \times 0.098 \div 4 = 0.00754$ square inches

Main Jet D	Jet Area (1)	Metering Rod Step	M. Rod Area (1)
0.086"	0.00581	0.037"	0.00108
0.089"	0.00622	0.042"	0.00139
0.092"	0.00665	0.047"	0.00174
0.095"	0.00709	0.052"	0.0021
0.098"	0.00754	0.057"	0.00255
0.101"	0.00801		
0.104"	0.00850		
0.107"	0.00899		
0.110"	0.00950		

(1)—all areas are in square inches

In this style of fuel delivery system, the fuel flow through the jet is directly related to the available area. With a primary jet and a metering rod, the actual area, effective area available or net area, is the difference in area between the jet and the metering rod. So using the Edelbrock 600 cfm Performer AFB as our example, it uses 0.098" main jets (primary) and a 0.047" small or wide-open-throttle step metering rod, so the net area would 0.00754 – 0.00174 = 0.0058 (areas are from above chart). These jet sizes come from Edelbrock's spec sheet, so I'll reprint that info here.

Performer ARB

Carburetor	Engine	Part Number	Main Jet	Metering Rods
500 cfm	225–327	1403, 1404	0.089"	0.052"
600 cfm	302–400	1406	0.098"	0.047"
750 cfm	350–502	1407	0.113"	0.047"

The 600 cfm carb listed above is rated for 302 to 400 cubic-inch engines. So let's assume that our engine is about 20% larger—say 425 to 450 inches. While I might consider the larger 750 cfm carb for this larger engine, let's assume that I don't want to change. To get the extra fuel into the combustion chambers because of the larger displacement (more displacement equals more cfm, which means more fuel to keep the A/F ratio the same) means larger jets. As a general rule, in this type of calculation, I would recommend leaving the metering rod at 0.047"—seems to be the most common out-of-the-box—because it allows you 2 steps richer and 2-steps leaner. Roughly speaking, you want to increase the net area of the jet by about 20% because displacement is 20% larger. Calculated above as 0.00580 square-inches, you therefore want a net area of 0.00696 (1.20 x 0.0058) and then add in the metering rod size which you aren't changing at 0.00174 (0.00174 + 0.00696) = 0.00870 area for the new jet. From the above chart, the 0.104" jet is the closest one to this area.

Tip: Don't get too carried away with this net-area calculation. It is just a method for estimating a change if you have no dynamometer or in-car data to go on. Generally, if you increase the engine's displacement, you will also increase the airflow rating on the carburetor. This bigger carburetor tends to come with bigger jets, so use the Edelbrock data as guideline for each displacement. You can use the same basic method to estimate jet size based on estimated horsepower output. More horsepower output from the engine means it will need more fuel to keep the A/F ratio the same. A 20% power gain needs more fuel at the same displacement. Also do not forget that the carburetor has secondary jets also which should be increased and there is no metering rod in the secondary jet. Obviously, you could put a big change in the primary, or a big change in the secondary or you could do both—split about half and half. I would recommend that both primary and secondary jets be increased in size about the same amount compared to the Edelbrock baseline.

covered in Chapter 6. However, these two aspects are not completely independent, and there is a fine line between what is part of the disassembly process and what is an adjustment. Most of the adjustments require the engine to be running, which requires the carburetor to be fully assembled and to be installed onto the manifold (Chapter 5). I didn't want the basic adjustments to require the carburetor to be disassembled a second time. The exceptions are the jets, float setting and velocity valve modifications. These adjustments must be accomplished while the carburetor is apart.

Metering Rod Removal—You can remove the metering rods at any time but they must be removed before you remove the top of the carburetor. Each metering rod on the AFB and AVS is held in with one screw. Hold your finger on top of the cover plate as you un-thread the screw. Then remove the whole assembly—metering rod, piston, cover, screw and the step-up spring. The spring tends to stay in the well so you have to pick it out separately.

Float Setting—The typical AFB/AVS float setting is 7/16", which is measured with the air horn upside down (inverted) and taken between the air horn gasket and the top, outer end of the float. This description can be tricky because the air horn is inverted. This means that the top of the float is really the bottom or the part that is closest to the gasket. In some cases, you may want to try to push the float height by allowing the float to sit higher. This requires the actual setting (7/16") to be less—say 6/16"—because the float is inverted. On the AFB, AVS and TQ models, the float setting cannot be changed once the carburetor is assembled, so this decision must be made while the carb is still apart.

Velocity Valve Modification—The velocity valve is located over the secondary throttle blades in the AFB. It rolls open to allow more air to flow to the engine as the airflow demand of the engine pulls the velocity valve open. Obviously the actual secondary throttle blades have to be open first. The basic speed and timing of this secondary opening is controlled by the counterweight on each end of the valve. To change the timing and speed of this opening, you must cut material off these counterweights. Similar to the jets, this change cannot be done once the carburetor is assembled. You want to modify the weights before you start the reassembly process. How much to cut off the weights? This is done by trial and error! The standard counterweight is probably good for a 12- to 13-second ET at the drag strip. To get the car down to the 11s, you might try cutting off half the weight. This is shown in photo number 4 on page 48—use the full amount shown for the 10-second bracket. Make sure you have a spare velocity valve before you perform this modification.

Note: Some of the production AFBs used a different shape counterweight. Since these carbs are no longer serviced, this part could be impossible to find. I think they were only used on smaller cfm versions that were designed for smaller displacements and lower horsepower ratings.

AFB

Disassembly

The following will be a step-by-step discussion covering how to disassemble the AFB carburetor. Although the AVS is similar, I'll discuss that separately in the next section, followed by the Thermo-Quad.

Caution: Do not discard any carburetor gaskets until you put the carb back together. Use the old gaskets to select your new gaskets since most rebuild kits have several possible selections.

1. Tip the carburetor upside down and drain the gasoline outside and away from any flame or heat source.

2. Set the carburetor on a carburetor work stand (you can get these from Moroso) or use two blocks of 2 x 4—one front and one rear. This tool is used to protect the throttle blades and linkage and provide a sturdy work platform. With the carburetor raised, the throttles can be opened freely which will be helpful in many steps.

3. Remove the fuel line fitting. This may require clamping the carburetor to the workbench, or while it is still bolted to the intake manifold. Once the fitting is removed, remove the filter screen.

Note: Screen is normally inserted into the backside of the fitting. The screen should come out with the fitting.

4. Remove the small clip that attaches the fast idle connector rod to the choke lever. Pull the rod out of the lever.

5. Rotate the rod until the tangs on the lower end allow it to be disengaged from the fast idle cam. Note or mark which hole it came out of, if there is more than one hole available.

6. Remove the choke diaphragm rod—small clip from end and rotate hooked end out of slide.

Note: Some rods may use 2 clips.

7. Remove the accelerator pump linkage. Remove

the small clip that holds the rod to the actuating lever. Note (mark) which hole the rod is in (the middle hole is most common).

8. Lower end of pump rod may be held in by a small clip or by the housing proximity. Remove the clip or rotate the rod/linkage so it can be removed from the throttle linkage.

9. Remove the screw (which acts as a pivot) that holds the pump's actuating lever in place and remove lever. First, note the small spring around head of the screw and how it is anchored on each end. Second, note the direction of the S-clip that attaches the lever to the actual accelerator pump. The top of the S-link should point away from the center of the carburetor (looks like an S not a Z when viewed from front of carburetor). Check sequence in which the linkage is removed—bowl vent on top, then pump link and S-link. Bowl vent and small spring are not used on all versions of AFB carburetor.

10. If not done earlier, remove the screws (one per side) that hold the metering rod cover plates while holding the cover plate down with your finger to prevent the piston and metering rod assembly from flying out. Lift off the plates and screws and slide the metering rods and pistons up and out of the air horn. Remove the step-up piston springs.

11. Remove the 8 or 10 screws that attach the air horn to the main body. The Edelbrock uses 8 Torx-head screws and older production AFBs use 10 standard screws. With 10 screws, 2 are longer than the others—one (1 1/2") is at center-rear and the other (1") is at center-front.
Note: Eight equal-length screws are 3/4" long.

12. Lift the air horn straight up and away from the main body. Use caution so as not to damage the floats. Gasket has to go with the air horn (top) and it may stick to the lower housing. Disconnect from lower housing by sliding a thin knife along surface. Be patient and don't hurry.

13. Remove the accelerator pump, the plunger and the lower spring from the pump cylinder.

Disassembling Air Horn—Flip the air horn upside down and place on bench.

14. Remove the float pins (left and right). Then lift the float up and out. Mark the pump side (left) float.

Tip: When upside down, it is hard to figure out left and right, so find the accelerator pump well and use it for side-to-side identification.

15. Remove the two needle check valves from their respective seats. Don't mix—keep left on left side, right on right side.

16. Using a wide-blade screwdriver, remove needle valve seats. Be sure that each needle valve is returned with its original seat at reassembly if new needle and seats aren't used. Place in separate plastic bags and label.

17. Carefully remove air horn gasket. Once removed, select a new gasket from the rebuild kit and overlay to be sure they are the same. Then discard old gasket.

18. Place the accelerator pump plunger, if it is going to be reused, in a jar of clean mineral spirits to prevent the seal from drying out.

Main Body Disassembly—Set the main body (lower section or bottom) on bench in normal orientation.

19. Remove and mark the left and right idle-mixture screws from front of carburetor. Mixture screws come with a small coil spring. Inspect the tapered ends (the pointed part) and if they have been damaged (grooved or bent) then find replacements.

20. Generally, the actual accelerator pump will stay with the air horn but if it stayed in the bottom, remove the piston/pump and then lift out the accelerator pump spring.

21. Remove the screws (2) that attach the accelerator pump shooter (nozzles) to the main body. Lift up the nozzle and gasket. Gasket tends to stick and may stay in housing. Remove gasket and use to select new gasket from rebuild kit. Then discard old gasket.

22. Next, flip the main body upside down and catch the discharge check needle or ball (and spring) as it drops out from the discharge passage.

23. Use a wide-blade screwdriver and remove the primary metering jets and the secondary metering jets. It is recommended that the primary jets be kept separate so they can be reinstalled in their respective locations—primary, secondary, left and right.

24. Remove the two screws that hold each primary booster venturi in place (one on each side). Lift each venturi straight up and away from body. Left and right venturi are not interchangeable.
Note: The brass air bleed tube on top of venturi housing should be to the rear (toward secondaries) on each side. Gaskets tend to stick to housing.

25. Remove the two screws that attach the large secondary venturi (left and right) to the main body and lift straight up and away from body. Note orientation: the key is the brass tube only goes into housing one way.

26. Remove the air valve. Note orientation.

27. If the carburetor is equipped with an electric choke, on the front of the right-side of the main body are 3 screws that hold choke housing. Note (mark) the orientation of the black plastic housing relative to the aluminum cast mount so the choke setting may be easily reset.
Note: There are 2 screws that hold the vacuum diaphragm to the housing in the production-style choke units.

28. Remove the backing gasket once the black plastic housing is removed.

29. Remove the three screws (2 internal under the black plastic housing and 1 external) that hold the choke housing to the main body.

30. It is usually not advisable and not recommended to remove the throttle blades and throttle shafts unless wear or damage requires new parts. At that point consider a new carburetor.
Note: If required, remember that the screws that attach the throttle blades to the throttle shaft are staked on the opposite side, so care should be used in removal so you do not break the screws in the throttle shaft. Remove the staked portion of the screws with a file prior to attempting removal.

31. Clean main body and remove any leftover gasket.

AFB Reassembly

To reassemble the carburetor, reverse the disassembly procedure. In many of the disassembly steps you marked various parts to help in this process when fitting the pieces back together. Be gentle! If it doesn't slide easily into place, something may be wrong. Correct before proceeding! Try not to force anything. Remember: Keep the old gaskets and use to identify the new ones.

1. Once cleaned, you are ready to reassemble.

2. Set aside the old gaskets especially the nozzle gasket and the venturi gaskets.
Note: use old gaskets to select new gaskets from rebuild kit.

3. Check throttle blade position. With the bottom housing upside down, fully open the throttle lever(s). The lever should hit its stop and the throttle blades should both be at 90 degrees to the carburetor base.

4. With the air horn inverted, install air-horn-to-main-body gasket in its proper position on the air horn.

5. Install the seat and tighten securely. Then install the needle.

6. Install the floats. With air horn upside down, slide right and left floats into the proper position and then install the float fulcrum (pivot) pins. Be sure that the marked float from disassembly procedure is installed in its original position. The floats should not be reversed.

7. Check float alignment. With air horn upside down, and with the floats installed, sight down the outside edge of each float (outside edge in each case). The outward side of the float and the outer edge of the air horn casting should be parallel (2 straight lines at equal distances).

8. Set float height. With the air horn upside down, the air horn gasket installed and the needle and seats installed, check the float height—outer end of the float to air horn gasket. Use 9/32" if no spec sheet info is available. Optional: use special float gauge, or steel scale or 9/32" drill.
Note: Any special float setting should be done at this time—see pages 44 and 56.

9. Check float drop. With the air horn right-side up or in its normal position, measure the float drop—distance from the top of the float's outer end up to the air horn gasket. It should measure 3/4".

10. Install the idle mixture screws and springs. Set idle mixture screws by adjusting the screws lightly against their seats (all the way in) and then back off 1 1/2 turns as an approximate initial adjustment.

This is best done using your fingers but otherwise use a very small screwdriver.

11. Install main jets and tighten securely. Install secondary jets and tighten securely. On production AFBs, install the accelerator pump check valve and tighten securely.
Note: If jets are being revised or changed for special reasons, see section at beginning of this chapter on how to estimate/calculate.

12. Install velocity valve. Cut weights prior to installation as required. If the velocity valve counterweights are going to be lightened or machined for quicker opening, do this operation prior to installation of the velocity valve. See section on velocity valve modification at the beginning of this chapter. Also see Chapter 6.

13. Install clusters and gaskets—both primary and secondary. May be easier to install gasket first.

14. After secondary clusters are in place, check velocity valve for ease of operation—no interference.

15. Install small accelerator pump check valve, point down. Install accelerator pump nozzle—two screws and gasket. Nozzles should point at primary throttle bores.

16. Install the 8 or 10 air horn screws. Edelbrock uses 8 Torx-head screws, all the same length. Production AFB uses 10 screws, the two longer screws—install in holes at the air cleaner mounting surface. The 1" screw goes at the front and the 1 1/2" screw is at the rear.
Note: The remaining 8 screws are 3/4" long.

17. Check accelerator pump travel. With the throttle fully closed, measure the distance from the top of the air horn to the top of the accelerator pump's plunger shaft using a steel scale.

18. Check secondary throttle opening. With the primary and secondary throttle blades fully closed, it should be possible to insert a 0.020" wire gauge or 0.020" feeler gauge between the positive closing shoes on the secondary throttle levers.

19. Check primary throttle stop. Adjust the primary idle screw adjustment to allow the primary throttle blades to close fully.

20. Bowl vent. If so equipped (production only). With the throttle blades fully closed, insert a 5/32" drill between the top of the air horn and the valve seal. The seal has a point in the center so measure to the outer lip.

21. Install the metering rods. The step-up piston, the metering rod and the step-up spring. Without specific information, use a 0.047" small step first since it allows several steps in both directions once you get into the fine tuning/adjustment phase. Push down gently on the piston and rod assembly until flush with housing and slide cover onto top and install screw and tighten.
Tip: You may have to wiggle the top of the assembly to get the rod centered so it drops into place.
Note: Remember that 2-step or 3-step metering rods must be used in sets—metering rods, main jet and cover.

AFB DISASSEMBLY

1. The metering rods can be removed first, if desired. Hold your finger on top of the cover plate as you loosen the screw.

2. Lift the metering rod assembly out of its well, and then lift out the step-up spring because it doesn't come out with the other parts.

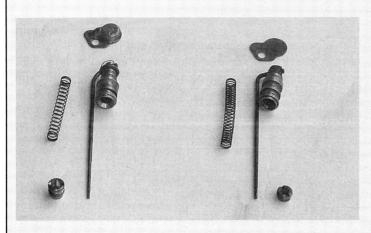

3. The pieces on the left are 3-step metering rod hardware and the ones on the right are 2-step. Keep all the metering rod parts together.

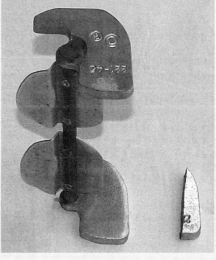

4. On the AFB carburetors, it is very important to decide if you are going to modify the velocity valve because it must be done while the carburetor is apart. This is a standard AFB velocity valve and the chunk to the right is the type of modification that is desired for very quick drag race cars—maybe 10 seconds. The same amount of material should be removed from each counterweight.

5. This is the chuck sitting in its normal position on the counterweight. Be sure to have spare velocity valves if you plan on modifications of this kind.

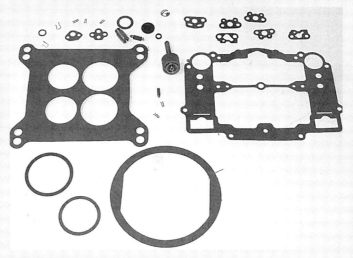

6. Before you start to actually disassemble the carburetor, be sure that you have a rebuild kit with lots of gaskets and small parts.

CARB REBUILDING

8. Use a large wrench to remove the fuel inlet fitting. Be sure to clamp the carburetor securely before attempting.

7. This is the top view of the AFB. Find the eight screws that hold the top on and then find the two extra—for 10 total. The Edelbrock AFB uses eight Torx screws while the production AFB uses 10. The extra two are—center top recessed below the air cleaner mounting surface and just below the PCV fitting. The second one is at the bottom center just outside the air cleaner's mounting surface.

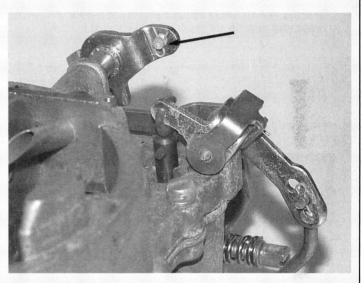

9. The inlet screen is usually located on the back side of the fitting. If it doesn't come out with the fitting, check inside the inlet and remove with needle-nose pliers if required.

10. Take the clip off the upper-end of the link for the fast idle connector rod.

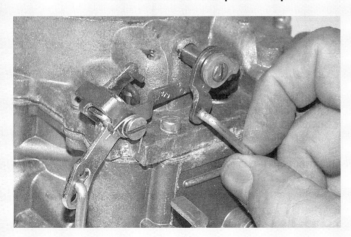

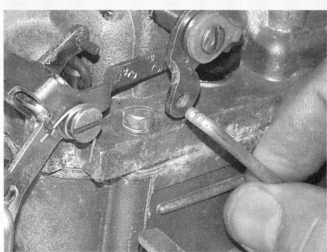

11. Pull the link out of the arm.

12. Once the link is free from pivot, you can rotate rearward.

49

13. Once rotated rearward, the tangs on the rod will line-up with the slot in the arm and the link can be removed.

14. Remove the clip from the end of the choke diaphragm rod using needle-nose pliers.

15. Remove one end of the rod from the diaphragm and rotate the hooked end out of the slide.

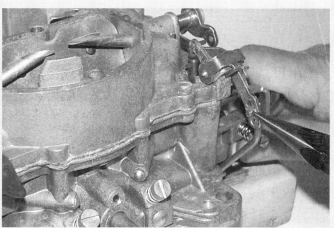

16. Pull the clip off the upper end of the accelerator pump linkage.

17. Remove the upper end of the link from the arm and rotate it so the lower end can be removed.

18. The accelerator pump arm is now free and held in by only the center pivot screw.

CARB REBUILDING

19. Remove the center pivot screw—note that there is a small spring under the arm and how it is installed before you remove the pivot. Note: The end of the spring is just to the left of the screwdriver blade.

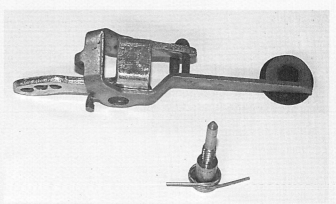

20. The parts to the accelerator pump linkage—the bowl vent arm sits over the accelerator pump arm and the spring sits around the head of the pivot. Note the shoulder on the pivot screw pilots into the linkage arms. Don't forget to remove the S-link in the top end of the pump itself.

21. If not done earlier, remove the screw that holds the metering rod cover in place. Hold cover with your finger until screw is out.

22. Lift the metering rod assembly out—which includes the rod, piston and spring clip.

23. Then lift out the small step-up spring.

24. Remove the 8 (Edelbrock) or 10 (production) screws. The Edelbrock carb uses Torx screws. Lift air horn straight up once free.

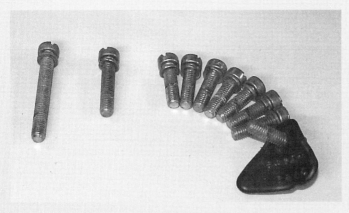

25. The production AFB uses 10 screws—8 of equal length, one 1" long and one 1 1/2" long. The Edelbrock uses 8 equal length screws. I leave the tag around the one screw so I don't forget it or lose it.

26. With the air horn and S-link removed, the accelerator pump will stay in the bottom housing. Lift it out.

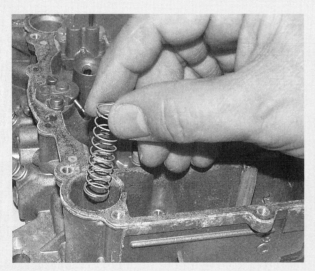

27. Then lift out the spring that sits below the pump.

28. With the air horn (top) upside down, the floats are now accessible. You remove the floats before the gasket.

29. Slide the pin out of the pivot.

30. Once the pin is out, the float can be removed.

CARB REBUILDING

31. Remember to mark at least one of the floats—left or pump side.

32. With the float removed from the air horn, the needle and seat are now accessible.

33. First remove the needle.

34. Then using a large-bladed screwdriver, remove the seat.

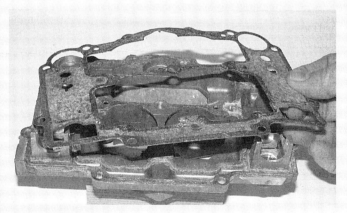

35. Carefully lift up the carburetor gasket.

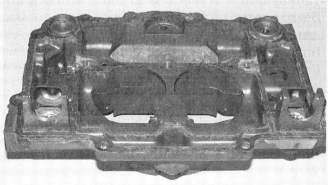

36. The air horn is now bare (upside down in photo)—set aside.

53

37. Remove the 2 screws that hold the accelerator pump shooter. Then lift it out.

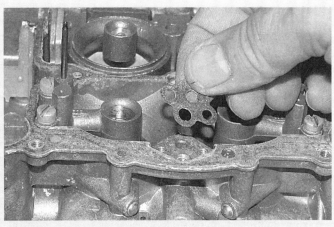

38. Find the small gasket that goes between the shooter and the housing. Sometimes it comes out with the shooter and sometimes it stays in the housing.

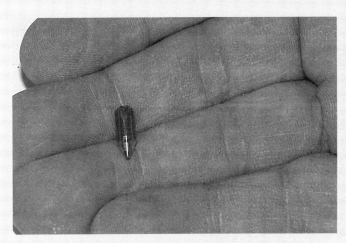

39. Flip the housing upside down and catch the small check valve located below the pump shooter and gasket.

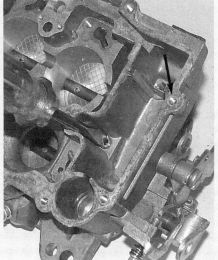

40. Using the thick, wide-blade screwdriver, remove the main jets and the secondary jets. The secondary jet is located above and to the right of the screwdriver blade, which is in the primary.

41. Remove the 2 screws that hold in the primary booster and then lift straight up. Note the gasket between the booster and the housing. Boosters must go back in on the same side as original so mark "pump side" or "left."

42. Remove the 2 screws that hold the secondary boosters and lift out. These must also go back in the original locations. Mark "pump side" or "left."

Carb Rebuilding

43. The primaries are toward the bottom and the secondaries toward the top. The AFB velocity valve is sitting above the secondary throttle blades.

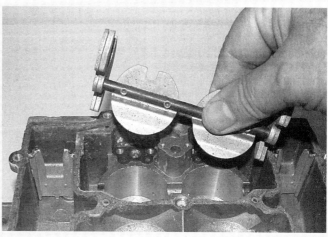

44. Lift the velocity valve out of its well, straight up until it is clear of the housing.

45. This is the velocity valve. There are several different shapes used. This one came from small cfm production carb.

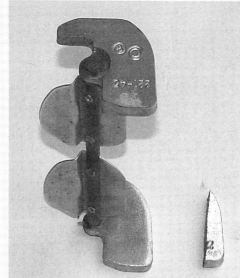

46. This is another style of velocity valve from a larger cfm production AFB. The chuck to the right is the amount that you might cut off of the counterweight for serious drag racing.

47. Remove any choke mechanism or diaphragm. It is typically mounted on the right side, toward the front.

AFB REASSEMBLY

1. Be sure the secondary linkage shoes have clearance between each other as they open fully.

2. Check the fully open throttle position.

3. Install carb gasket first. Then select the new seat and new seat gasket.

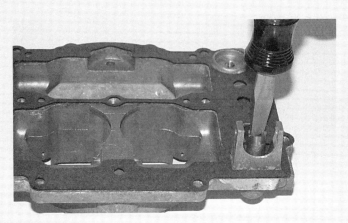

4. Install the seat and gasket, then tighten securely.

5. With the carb gasket installed and the air horn upside down and the floats installed, measure the float height between the bottom of the float and the top of the gasket as positioned. You can use the proper size drill or a steel scale.

Carb Rebuilding

6. If the float setting needs to be changed, you remove the pin and the float and you bend the tang that is located between the two flange pivots in the lower left of the photo.

7. Use a pair of pliers to bend the tang.

8. Check the alignment of the floats by comparing the outside wall of the float to the straight side on the outside of the housing.

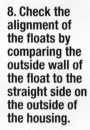

9. Check the float height (drop) using a steel scale.

10. Install the idle mixture screws—remember the spring.

11. Install the main jets and tighten.

12. Be sure that both the primary and secondary jet is installed on each side and secure.

13. Install the AFB velocity valve to the right of center—its shaft is vertical in photo. Once into its well, rotate the counterweight forward to be sure that it works freely.

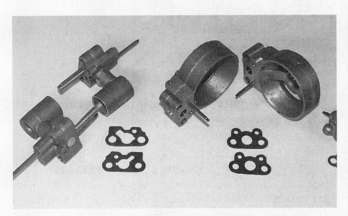
14. Gather the primary and secondary clusters and the gaskets. Use the old gaskets to select new ones if you have several styles.

15. Install the gasket and then the secondary cluster. Install the screws and tighten. Remember that the cluster must go in the same location as the original, pump side first.

16. Install the gasket onto the primary cluster and install the assembly into the housing. Install the screws and tighten. Remember that they must be installed in the same position as the original, pump side first. Install accelerator pump spring and pump in well at bottom (arrow).

17. With the secondary clusters installed, make another test run by rolling the velocity valve again, to be sure that it clears everything—it's below the secondary clusters, to the left in the photo. The arrows show the tips of valves.

18. Flip the air horn over and install onto the main housing. It must drop straight down into position. Push the accelerator pump shaft down to be sure that it is free. Make sure that the two parts are flush and that the gasket hasn't moved, and then install the 8 or 10 screws and tighten.

19. Insert the screen into the fuel inlet filter, install assembly into the carburetor and tighten. Clamp securely.

20. Reassemble the various links in reverse order.

AVS

Disassembly

The following will be a step-by-step discussion covering how to disassemble the AVS carburetor.

Caution: Do not discard any carburetor gasket until you put the carb back together. Use the old gaskets to select your new gaskets since most rebuild kits have several possible selections.

1. Tip the carb over and drain the gasoline outside and away from any flame or heat source.

2. Set the carburetor on a work stand (available from Moroso) or two blocks of 2 x 4 wood—one front and one rear. This tool is used to protect the throttle blades and linkage and provide a sturdy work platform.

3. Remove the fuel line fitting. This may require the carburetor to be clamped securely to the workbench or still bolted to the intake manifold. Once fitting is removed, remove the filter screen. Clean if required.

4. Remove the small clip that attaches the fast idle connector rod to the choke lever. Pull the rod out of the lever.

5. Rotate the rod until the tangs on the lower end allow the rod to be disengaged from the fast idle cam or remove clip on lower rod end, depending upon style. Note or mark the hole that it came out of if more than one hole is available.

6. Remove the accelerator pump linkage. Remove the small clip that holds the upper end of rod to the actuating lever.
Note: Mark the hole where the rod was located for reassembly.

7. Lower end of rod may be held in by a small clip or by tangs on the lower end of the rod. Remove the clip or rotate the rod so it can be removed from the throttle linkage.

8. Remove the shoulder screw (acts as pivot) and spring that holds the pump's actuating lever and bowl vent arm in place and remove lever and arm. Note direction of the S link that attaches the lever to the actual accelerator pump. Top of S-link should point away from the center of the carburetor (looks like an S not a Z when viewed from front of carburetor).

9. On production AVS carbs, remove the vacuum hose between the throttle body (bottom) and the vacuum diaphragm (part of choke located on right-front of carburetor). Electric choke versions are discussed in step 30 on the next page.

10. Remove the small clip from the choke operating link and disengage link from the diaphragm plunger and the choke lever.

11. Remove the vacuum diaphragm and bracket assembly from side of carburetor and place in a safe place.

12. If not done earlier, remove the screws (one per side) that hold the metering rod cover plates while holding the cover plate down with your finger to prevent the piston and rod from flying out. Lift off the plates and screws and slide the metering rods, pistons out of the air horn. Remove the step-up piston springs.

13. Remove the 8 screws that attach the air horn to the main body. The Edelbrock uses 8 Torx-head screws. All AVS screws are the same length.

14. Lift the air horn straight up and away from the main body. Use caution so as not to damage the floats. The gasket tends to stick, so use care. It should go with the air horn.

15. Remove the accelerator pump, the plunger and the lower spring from the pump cylinder.

Disassembling Air Horn—Flip the air horn upside down and place on bench.

16. Remove the float pins (left and right). Then lift the float up and out. Mark the pump-side float. Floats need to be reinstalled in original locations.

17. Remove the two needle check valves from their respective seats. Don't mix—keep left on left side, right on right side.

18. Using a wide-blade screwdriver, remove needle valve seats. Be sure that the each needle valve is returned with its original seat at reassembly.

19. Slide accelerator pump plunger from air horn, if pump stayed with the top.

20. Place the accelerator pump plunger, if it is going to be reused, in a jar of clean mineral spirits to prevent the seal from drying out.

21. Remove air horn gasket carefully. Set aside the bare air horn.

Main Body Disassembly—Set the main body (lower section or bottom) on the bench in normal orientation.

22. Remove the center idle-mixture screw from front of carburetor.
   Note: There are 2 mixture screws on Edelbrock AVS. Mixture screws come with a small coil spring. The production AVS carburetors use one center mixture screw and it has a left-hand thread. The Edelbrock AVS versions use two mixture screws and they have right-hand threads like the AFB. Inspect the tapered end (the pointed part) of the screw and if it has been damaged (grooved or bent) then find replacement. Some production carbs may have a plastic cap over the head of the mixture screw— remove cap first.

23. Generally, the actual accelerator pump plunger will stay with the air horn, but if it stayed in the bottom, remove the piston and then lift out the accelerator pump spring.

24. Remove the screws (2) that attach the accelerator pump shooter (nozzles) to the main body. Lift up the nozzles and gasket. Gasket may stick. Use old gasket to select new gasket from rebuild kit and then discard the old gasket.

25. Next flip the main body upside-down and catch the discharge check needle or ball (and spring) as it drops out from the discharge passage.

26. Remove the two screws that hold the primary booster venturis in place (one on each side). Lift each venturi straight up and away from body. Left and right venturis are not interchangeable. **Note:** brass air bleed tube on top of venturi housing should be to the rear on each side.

27. Use a wide-blade screwdriver (or the special jet tool) and remove the primary metering jets and the secondary metering jets. It is recommended that the primary jets be kept separate so they can be reinstalled in their respective locations. Edelbrock uses the same jet on the left and right side, but this is not true on all production carburetors. The 3-step primary jet common on production carbs is easy to identify from the shorter secondary jets but on Edelbrock carbs, the primary and secondary jets are the same in appearance but not always in diameter size.

28. On the Edelbrock AVS versions, remove the two screws that attach the large secondary venturi (left and right) to main body and lift straight up and away from body. On the production AVS carburetors, there is no secondary venturi. Note that the production versions of the AVS use nozzle bars in the secondary—one on each side. Do not try to remove the nozzle bars.

29. Using a wide-blade screwdriver or special tool, remove the accelerator pump intake check valve located inside the left-side (accelerator pump side) fuel bowl, next to the accelerator pump cylinder. Production carbs only.

30. If the carburetor is equipped with an electric choke (Edelbrock versions), on the front of the right-side of the main body, you can remove—3 screws hold choke housing. Note (mark) the orientation of the black plastic housing relative to the aluminum cast-mount, so the choke setting may be easily reset. On production carbs, the vacuum diaphragm is secured by 2 screws.

31. Remove the backing gasket once the black plastic housing is removed.

32. Remove the three screws (2 internal under the black plastic housing and 1 is external) that hold the choke housing to the main body.

33. It is usually not advisable and not recommended to remove the throttle blades and throttle shafts unless wear or damage requires new parts. At that point consider a new carburetor.
   Note: If required, remember that the screws that attach the throttle blades to the throttle shaft are staked on the opposite side care should be used in removal so you do not break the screws in the throttle shaft. Remove the staked portion of the screws with a file.

34. Clean main body and remove any leftover gasket.

AVS Reassembly

To reassemble the carburetor, reverse the disassembly procedure. In many of the disassembly steps you marked various parts to help in this process. Be gentle! If it doesn't slide easily into place, something may be wrong. Correct before proceeding. Try not to force anything.

1. Once cleaned, you are ready to reassemble the carb.

2. Use the old gaskets for the nozzle and venturi to select the new gaskets from the rebuild kit. Then discard the old gaskets.

3. Check throttle blade position. With the bottom upside down, fully open the throttle lever. The lever should hit its stop and the throttle blades should both be at 90° to the carburetor base.

4. With the air horn inverted, install air-to-main body gasket in its proper position on the air horn.

5. Install each seat and tighten securely, then install each needle.

6. Install the floats. With air horn upside down, slide right and left (accelerator pump side) floats into the proper position and then install the float fulcrum (pivot) pins. Be sure that the marked float (pump-side) from disassembly procedure is installed in its original position. The floats should not be reversed.

7. Check float alignment. With air horn upside down, and with the floats installed, sight down the side of each float (outside edge in each case). The outward side of the float and the outer edge of the air horn casting should be parallel.

8. Set float height. With the air horn upside down, the air horn gasket installed and the needle and seats installed, measure the float height—outer end of the float to air horn gasket. Use 9/32" if no spec sheet info available. Use special float gauge, or steel scale or 9/32" drill.
Note: Any special float setting should be done at this time—see beginning of chapter.

9. Check float drop—with the air horn right-side up or its normal position, measure the float drop—distance from the top of the float's outer end up to the air horn gasket. It should measure 3/4".

10. Install the idle mixture screw (center) and spring. The single production screw is a left-hand thread. The Edelbrock AVS versions use 2 right-hand screws. Set idle mixture screws by adjusting the screw lightly against its seat (all the way in) and then back off 1 1/2 turns as an approximate initial adjustment. This is best done using your fingers but otherwise use a very small screwdriver.
Note: The production mixture screw is a left-hand thread.

11. Install main jets and tighten securely. Install secondary jets and tighten securely. On production carbs, install the intake check, in fuel bowl next to accelerator pump cylinder.
Note: if jets are being revised or changed for special reasons, see section on page 43 in this chapter on how to estimate or calculate the size.

12. Install primary clusters and gaskets. It may be easier to install the gasket first. On Edelbrock versions, install secondary clusters and gaskets.

13. Install the accelerator pump intake check ball on production versions (if used).

14. Install the small discharge check needle, point down, into the accelerator pump channel.

15. Install the accelerator pump nozzle gasket. Install the nozzle and the 2 screws and tighten.

16. Place the accelerator pump into position in the pump well with the spring below it.

17. Install the air horn onto the main body, straight down.

18. Guide the floats past the housing gently. Be sure that the accelerator pump shaft slides through its hole in the air horn—see photo. Double check that the air horn is flush with the main body.

19. Install the 8 air horn attaching screws. Edelbrock uses 8 Torx-head screws, all the same length. Tighten securely.
Note: If there are 2 longer screws, they go in the dash-pot mounting bracket. Production carbs, manual transmission are only versions to use dashpot.
Tip: Tighten 1 screw on each side (left and right) first and then tighten rest. Retighten all 8.

20. Install accelerator pump linkage. Edelbrock and production versions are somewhat different. Be sure pivot screw pilots on levers and spring is installed on production versions.

21. Check out accelerator pump travel. With the throttle fully closed, measure the distance from the top of the air horn to the top of the accelerator pump's plunger shaft using a steel scale.

22. Check secondary throttle opening. With the primary and secondary throttle blades fully closed, it should be possible to insert a 0.020" wire gauge

or 0.020" feeler gauge between the positive closing shoes on the secondary throttle levers.

23. Check fast idle cam and idle adjustment. Is it functioning? It is easier to check fast idle or throttle stop with the carb off the engine.

24. Check primary throttle stop. Adjust the primary idle screw adjustment to allow the primary throttle blades to close fully. Sometimes idle adjustment holds primary blades slightly open and this will cause problems at tune-up time.

25. Bowl vent (production only if so equipped). With the throttle blades fully closed, insert a 5/32" drill between the top of the air horn and the valve. This feature is not used on Edelbrock versions.

26. Install the metering rods—the step-up piston, the metering rod and the step-up spring. Without specific information use a 0.047" small step first since it allows you several steps in both directions once you get into the fine tuning/adjustment phase. Push down gently on the piston and rod assembly until flush with housing and slide cover onto top and install screw and tighten.

Note: Remember that 2-step or 3-step metering rods must be used in sets—metering rods, main jet, cover.

Tip: You may have to wiggle the top of the assembly to get the rod centered so it drops into place.

27. The production AVS uses a vacuum diaphragm choke while the Edelbrock uses an electric choke. Be sure that they are hooked up properly.

28. Adjust secondary air valve. Loosen the lock screw, smaller one to the outer edge. Allow the air valve to drop down—loosen and it should hang in the vertical, wide-open position.

29. Turn the larger screw and it will tighten the spring putting tension on the air door until it is just closed and then go 1 1/2" additional turns. This procedure is also given as 2 turns total measured from the vertical position.

Note: It may help to put tape on one side of the screwdriver blade to help keep track of full turns and 1/2 turns in this step.

30. Hold this position with one screwdriver and tighten the lock screw with a second (smaller) screwdriver.

31. Slide the fuel inlet screen into the fuel line fitting until seated and install fitting.

AVS DISASSEMBLY

1. The top view of the Edelbrock AVS. Note the 8 Torx screws that hold the carburetor together. Primaries are toward the top.

2. Use a large wrench to remove the fuel inlet fitting. Be sure to clamp the carburetor securely.

3. The fitting has a screen on the inside. It actually fits inside the fitting itself.

4. Remove the clip from the upper end of the fast idle connector rod.

5. Remove the clip from the lower end.

CARB REBUILDING

6. Remove the link.

7. Remove the clip from the upper end of the accelerator pump linkage rod.

8. Pull the rod out of the arm and allow it to drop down, then remove from the lower part of the linkage.

9. Remove the center pivot of the accelerator pump arm. Note that there is a spring centered on the pivot screw and note its anchor points—one is just to the left of the screwdriver blade.

10. The pivot and the spring removed from the arm mechanism. Note the large pilot on the head of the screw. The accelerator pump arm centers on this diameter.

11. The bowl vent arm sits over an accelerator pump arm. Here it has been removed and you can see the S-link top pointing to the right.

12. Remove the accelerator pump arm and S-link (arrow).

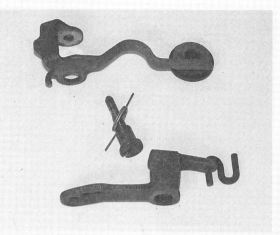
13. The bowl vent arm (top), and the accelerator pump arm and its S-link (lower) are both held in with the center pivot and spring (middle).

14. Remove the clip from one end of the linkage and remove link from the vacuum diaphragm.

15. Remove the clip from the opposite end of the link and then remove the link.

16. Remove the screw from the metering rod cover plate. Hold the cover plate down with your finger during this operation.

17. Lift the metering rod and piston assembly straight up.

Carb Rebuilding

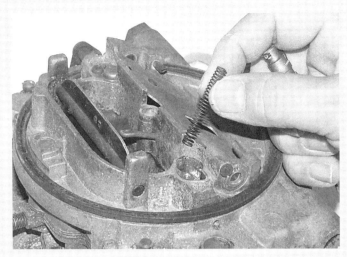

18. Lift out the step-up spring that is left in the well.

19. Remove the 8 screws that hold the carburetor together. The Edelbrock version uses Torx screws. On the production version, one screw, usually on the right front, has a tag on it.

20. Lift the air horn straight up. Remember that there is a float on each side. The accelerator pump itself may stay with the bottom. The carb gasket must come off with the top air horn.

21. To disassemble the air horn, flip it over and set on the bench.

22. Pull the float pin out of the pivot.

23. Mark the float so you can put it back in the same location. Then remove the float.

Rebuild & Powertune Carter/Edelbrock Carburetors

24. Remove the needle.

25. Using a large-bladed screwdriver, remove the seat.

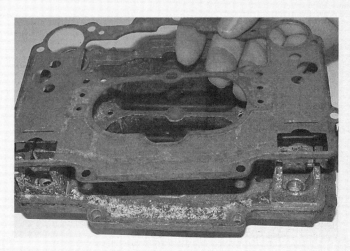

26. Gently lift up the gasket and set aside.

27. The air horn is now bare (shown upside down).

28. Remove the center idle mixture screw on the production AVS. The Edelbrock AVS has 2. The single center screw has a left-hand thread. Edelbrock AVS screws are right-hand thread.

29. Each mixture screw has a small coil spring held under the head.

CARB REBUILDING

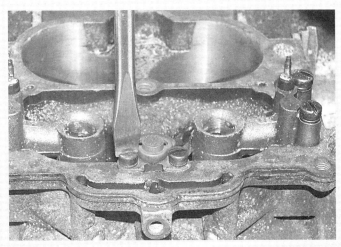

30. Remove the 2 screws that hold the accelerator pump nozzle in. Then lift up the nozzle and its gasket.

31. Then flip the carb body over and catch the tiny discharge check needle.

32. Remove the two screws that hold the primary booster venturi in place.

33. Lift the primary booster venturi out. Mark the pump side because booster positions can't be reversed.

34. If the booster gasket didn't come out with the booster, then remove the gasket gently, so it can be used to select the new gasket.

35. On the production AVS there are no secondary booster venturi—only the nozzle bars as shown—upper two secondary throttle bores—tubes that stick across the bores almost horizontally are nozzle bars.

36. Using a thick, wide-blade screwdriver, remove the primary jets.

37. Next remove the secondary jets.

38. In the production AVS, which uses the 3-step metering rods, the primary jet shown on the left is taller than the standard jet shown on the right. The right jet is used in both the primary and secondary locations on the Edelbrock AVS versions.

39. Using the wide-blade screwdriver, remove the accelerator pump intake check valve located on the left side of the housing, at the bottom of the accelerator pump well.

40. The tall primary metering jet is in the center, the short jet is to the right and the accelerator pump check valve is on the left—similar but unique. Bag and label when you remove them.

CARB REBUILDING

AVS REASSEMBLY

1. Check the wide-open throttle position—the throttle blades should be vertical.

2. Install the new main body gasket, then install the float. Check the float height using the proper sized drill between the bottom of the float and the top of the gasket.

3. You can also measure the float height using a steel scale.

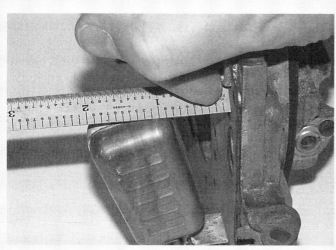

4. Tip the air horn back right-side up and measure the float drop.

5. Check the float alignment using the side of the float with the straight side of the housing (left side of photo).

6. The production primary cluster has a small gasket that seals between the cluster and the housing. Use the old gasket to select a new one. Note the small seal ring on the right. Note small gasket orientation.

7. The small ring or seal (arrow), installed but only pushed halfway into position. It should seat against the cluster housing.

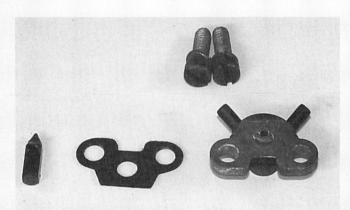

8. Gather the accelerator pump nozzle hardware—the discharge check needle is on the left.

9. Place the nozzle gasket on the nozzle or into position and then install the nozzle. Put the 2 screws into the nozzle and tighten. Note the new accelerator pump is sitting in position on the lower right.

10. Flip the air horn over (normal orientation) and lower it into position. Drop it straight down. Be sure that the accelerator pump shaft fits through the air horn. Push the pump down to be sure it is free.

11. Check alignment of the two parts and the gasket. Note the accelerator pump shaft sticking up in the upper right.

CARB REBUILDING

12. Check the clearance between the primary and secondary linkage shoes at part throttle opening. I'm using a steel scale here.

13. Check the wide-open-throttle stop—on the side of the housing.

14. Check the adjustable throttle stop—closed position and adjust so that the throttles are fully closed. Adjustment is made from the front of the carburetor on the left side or throttle side.

15. Adjust the fast-idle setting on the linkage. You want it to open the primary throttles slightly. This throttle adjustment will be fine-tuned once installed on the manifold. However, the adjusting screw is on the bottom of the linkage and is best accessed now so you can be familiar with its location.

16. Install the diaphragm or choke linkage.

17. Install the metering rod assemblies. The step-up spring goes in first. Place the cover on top of the metering rod assembly and wiggle down into position. Once aligned, push down gently and hold in position while you tighten the screw.

18. Check the bowl vent. With the throttle closed the height above the air horn can be measured with the proper size drill.

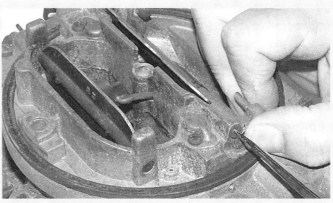

19. The air door on the AVS is spring loaded. There are two screws located on the left side of the carburetor. One is the lock screw and just holds the adjustment in place. The second screw is the actual spring adjustment. You loosen the outer screw but try to hold the inner screw so that it doesn't turn.

20. Once the outer screw is loose, the inner screw can be turned and the spring tensioned. A starting point is 2 1/2 turns. Define your own technique—measured from when the air horn is hanging vertical or from when it just closed.

21. Place the screen inside the fuel inlet fitting, the gasket on the outside and install into the air horn.

22. Tighten the fuel inlet fitting. Be sure to clamp the carb securely.

THERMO-QUAD
Disassembly

The production Thermo-Quad carburetors of this era have many vacuum lines and hoses. I would recommend marking the PCV vent, charcoal canister purge port, the EGR (exhaust gasket recirculation) and choke diaphragm. Use masking tape to label as required.

Caution: Do not discard any carburetor gasket until you put the carb back together. Use the old gaskets to select your new gaskets since most rebuild kits have several possible selections.

1. Tip the carb over and drain the gasoline outside and away from any flame or heat source.

2. Set the carburetor on a work stand (available from Moroso; their AFB stand works for the TQ) or two blocks of 2 x 4—one front and one rear. This tool is used to protect the throttle blades and linkage and provide a sturdy work platform. For general TQ identification, use the two tall and thin bowl vent tubes located at the rear of the carburetor behind the secondary air door as the '72-and-newer group. The '71 Thermo-Quad uses one large vertical vent-pipe, centered at the rear.

3. Remove the fuel line fitting. This may require the carburetor to be clamped securely to the workbench or still be bolted to the intake manifold. Once fitting is removed, remove the filter screen, if installed.

4. Remove the choke countershaft fast idle lever attaching screw (small 3/16 socket) while holding the lever in position. Remove lever from countershaft, and then swing fast idle connector up and forward until it can be disengaged from the fast idle operating lever.

5. Remove rod retainers and washer that holds the choke diaphragm connector rod to the choke vacuum diaphragm and air valve lever.

6. Remove the screws (one per side) that hold the metering rod cover plates. Note the tab or foot that sticks out on the left-side (pump-side) retainer. Lift off the plates and screws (one on each side) and set aside.

7. Remove screw that holds spring clip and piston-link assembly and remove piston, link and both metering rods. Use care because the metering rods can fall off the hanger once out of the housing.

Note: the dimples in the top of the hanger bar should face the choke (front).

8. Remove the step-up piston spring in center.

9. Remove the screw that attaches the accelerator pump shooter (nozzles) to the main body. Hold choke-blade open during this operation, if not in vertical position. Lift up the nozzle assembly and gasket. Compare to new gasket in rebuild kit and then discard the old gasket (at reassembly).

10. Invert the carburetor and remove the small discharge check needle (catch).

11. Remove the black plastic caps from the tops of the idle mixture screws using a pair of pliers. Mixture screws should now be visible.

12. Remove the accelerator pump linkage. Remove the small retaining clip on lower rod end and washer that holds the rod to the actuating lever (lower end of rod).

Note: Mark the hole where the rod was located for reassembly (upper end).

13. Upper end of rod is held in by bent hook in the end of the rod. Rotate the rod so it can be removed from the throttle linkage.

14. Remove the shoulder screw (acts as pivot) and spring that holds the pump's actuating lever in place and remove lever. Note direction of the S-clip that attaches the lever to the actual accelerator pump. The top of S-link should point away from the center of the carburetor (looks like an S not a Z when viewed from front of carburetor).

Tip: Leave S-clip in pump since it holds the accelerator pump up. It will be important at reassembly.

15. Remove the small clip from each end of the choke operating link and disengage link from the diaphragm plunger and the choke lever.

16. Optional: Remove the vacuum diaphragm and bracket assembly from right-side rear of carburetor and place in a safe place. See Throttle Body Disassembly, page 76.

17. Remove the 10 long screws that attach the air horn to the main body and the throttle body together (3-pieces). Two screws are located between the choke blade/valve and the center-wall of the bowl cover (also called the air horn). All screws are

fairly long and thin, and the same length.

18. Lift the air horn straight up and rotate to the rear. The rotated position allows the transfer rod to the throttle linkage to be unhooked. Then set the air horn away from the main body and flip over (upside down) to protect floats. Use caution during process so as not to damage the floats. There are two gaskets. Top gasket should go with the top. Try to remove this gasket in one piece.

Disassembling Air Horn—Flip the air horn upside down and place on bench, protect the floats.

19. Remove acceleration pump passage tube (plastic). This tube is about 4" long and very small. It must be reused. Use caution. It is easier to remove the gasket if this tube has previously been removed.

20. Remove air horn gasket. It is easier to remove the float pivots with the gasket removed. Use a thin, sharp knife and lots of patience. You tend to pry more than cut. Top gasket is a very close fit around many features.

21. Note the nozzle bars over the secondary openings—do not remove.

22. Remove the float pins (left and right). Then lift the float up and out. Mark the pump-side float. Use tape to ID if required. Floats need to be reinstalled in original locations.
Note: Use pump side for left-side reference. The floats are made of nitrophyl material (black).

23. Remove the two needle-check valves from their respective seats. Don't mix—keep left on left side, right on right side.

24. Using a wide-blade screwdriver, remove needle valve seats. Be sure that each needle valve is returned with its original seat at reassembly (if being reused).

25. Remove secondary metering jets. Use flats on side of tall secondary jets. Use 5/16" open-end wrench. Note how tall they are relative to the primary jets.

26. Remove accelerator pump rod S-link if not done earlier—pull out with pliers. To remove pump plunger assembly, use a small rod (or punch) placed on the upper end of the plunger shaft and tap lightly with a small hammer. Caution: do not damage plunger shaft hole in bowl cover.
Tip: This operation should be done with fingers under the lower portion of the pump cylinder so you can catch intake check seat, pump plunger and spring. Check seat will be damaged in this operation so always replace—install new check seat upon reassembly.
Tip: Catching parts in your hand is not easy, so provide a soft and safe place for them to land.

27. Place the accelerator pump plunger, if it is going to be reused, in a jar of clean mineral spirits to prevent the seal from drying out.

Main Body Disassembly (the center part)—This is the black phenolic (plastic) resin part. Set the main body (lower section or bottom) on bench in normal orientation. The bottom gasket tends to stick to this part.

28. Carefully remove the bottom gasket if it is attached to the main body.

29. Remove the primary O-ring seals (1 per side). They tend to stick up inside the well. Caution: Do not discard unless new O-rings are available in rebuild kit.

30. Remove the primary metering jets. Use a large-blade screwdriver or special tool.

31. Optional: Note the secondary throttle bore diffuser, which is a flat piece of sheet-metal that sits vertically across the secondary opening. You can remove or leave in place—do not force it.

32. Do not immerse main body in cleaning solvents for a prolonged period of time.

Throttle Body Disassembly (bottom)—An aluminum casting that holds the throttle blades.

33. Remove the gasket if not done above. Remove step-up actuating lever (in front section between the primaries) and pin.

34. Remove the choke diaphragm and bracket (right rear)

35. Remove plastic limiter caps from idle mixture screws, if not done earlier.

36. Remove the two idle-mixture screws from front of carburetor. Mixture screws (standard thread) come with a small coil spring. Inspect the

tapered end (the pointed part) and if it has been damaged (grooved or bent) then find replacement.

37. Note that the primary boosters are cast into the air horn and not removable as separate parts.

38. On the production TQ carburetors, there is no removable secondary venturi or cluster. The shape is cast into the main body plastic housing.

39. Remove the vacuum hose between the throttle body (bottom) and the vacuum diaphragm (part of choke located on right rear of carburetor). This can be removed earlier but better wrench access is provided once the main body is removed. Use a 5/16 socket.

40. Remove the idle-assist solenoid (California-only models)—1 screw, right front of base.

41. If the carburetor is equipped with an electric choke, on the rear of the right-side of the main body, you can remove—3 screws hold choke housing. Note or mark the orientation of the black plastic housing and hole relative to the aluminum cast mount, so the choke setting may be easily reset.

42. Remove the backing plate once the black plastic housing is removed.

43. Remove the two screws (1 is recessed and easy to miss) that hold the choke housing to the main body.

44. It is usually not advisable nor recommended to remove the throttle blades and throttle shafts unless wear or damage requires new parts. At that point consider a new carburetor.
Note: If required, remember that the screws that attach the throttle blades to the throttle shaft are staked on the opposite side. Care should be used in removal so you do not break the screws in the throttle shaft. Remove the staked portion of the screws with a file.

45. Clean main body and remove any leftover gasket.

TQ Reassembly

To reassemble the carburetor, reverse the disassembly procedure. In many of the disassembly steps you marked various parts to help in this process. Be gentle! If it doesn't slide easily into place, something may be wrong. Correct before proceeding! Try not to force anything.

Throttle Body Prep

1. Once cleaned, you are ready to reassemble. The first step is to prep the throttle body.

2. Be sure gasket surface is smooth and the old gasket has been completely removed.

3. Install the idle mixture screws (2) and springs. Set idle mixture screws (right-hand thread) by adjusting the screw lightly against its seat (all the way in) and then back off 1 1/2 turns as an approximate initial adjustment. This is best done using your fingers or a very small screwdriver.

4. Place lever and pin into position between the primaries.

5. Install the vacuum diaphragm on right rear of housing. Easier to do now than later.

6. Install idle assist solenoid on right front if being used (California carbs only).

7. Use the old, original gaskets to select the new gaskets from your rebuild kit. Then discard the old gaskets, especially the nozzle gasket and the venturi gaskets. Don't discard a part until its proper replacement has been selected.

Main Body Reassembly Prep (the black plastic part)

8. Be sure both gasket surfaces are smooth and the old gasket has been completely removed.

9. Install main jets and tighten securely.
Note: If jets are being revised or changed for special reasons, see section on page 43 of this chapter on how to estimate/calculate.

10. Install the diffuser. Be sure that it sits flush with the top surface.

11. Install the bottom gasket. Push on over the 4 locating dowels cast into the bottom surface of the main body (black plastic part).

12. Install the two O-rings, one on each side. Tap down to seat O-ring squarely in well.

13. Slide the fuel inlet screen into the fuel line fitting.

Air Horn Reassembly Prep

14. Be sure gasket surface is smooth, with the old gasket completely removed.

15. Install the accelerator pump plunger and spring. Place the intake check seat on smooth, flat surface. Push the plunger up the pump well so the shaft extends out the top and insert the S-link into the top of the plunger and release pressure so the S-link now holds pump in place. Locate the bottom of the pump well squarely over the check seat and push securely into place. Tap if required.

16. With the air horn inverted, install the inlet seat and tighten securely. Then install the needle.

17. Install the floats. With air horn upside down, slide right and left floats into the proper position and then install the float fulcrum (pivot) pins. Be sure that the marked float from disassembly procedure is installed in its original position. The floats should not be reversed. It may be easier to install the float pivots before the gasket is installed.

18. Install air horn to main body gasket in its proper position on the air horn. Push down around all of the special features—it is a very tight fit.

19. Set float height. With the air horn upside down, the air horn gasket installed and the needle and seats installed, check the float height—outer end of the float to air horn gasket. Use 29/32" if no spec sheet info available. Use special float gauge, or steel scale. Bend float arm to adjust. Use pliers and DO NOT bend against needle and seat.
Tip: For performance, set slightly high which means less or lower number. Measure from gasket to bottom of float, which is actually the top in the upside-down position.
Note: Any special float setting should be done at this time—see page 44 of this chapter.

20. Install secondary jets and tighten securely.
Note: If jets are being revised or changed for special reasons, see page 43 of this chapter on how to estimate/calculate. Remember to coordinate with any change in the main jets—see step 9.

21. Install the small plastic tube to the bottom of the accelerator pump cylinder and fitting at the bottom of the nozzle well.

22. With the air horn in its normal orientation, install the accelerator pump check valve. Then install the accelerator pump nozzle and gasket and screw and tighten securely.

Three-Piece Assembly
23. Place the throttle body on flat surface and locate the main body and gasket over it and push dowels into place. The two pieces should fit snuggly together—no gaps.

24. Place the air horn (top) and gasket onto the lower two pieces using care with the floats. It must be lowered straight down into the bowls. Remember to reinstall the transfer rod to the throttle linkage. Install rod in throttle linkage first and hold rod upwards. Rotate the air horn upwards and slide link/rod into place. Rotate air horn to normal position and lower into place.

25. Install the 10 air horn attaching screws, which are all the same length. Tighten securely.
Note: If there are 2 longer screws, they go in the dash pot mounting bracket. Production carbs, manual transmission only. Don't forget the 2 screws in the choke tower, behind the choke blade.

26. Check accelerator pump travel. With the throttle fully closed, measure the distance from the top of the air horn to the top of the accelerator pump's plunger shaft using a steel scale.

27. If not done earlier, install the accelerator pump check valve. Then install the acceleratotr pump nozzle and gasket using the screw.

28. Tighten the nozzle screw securely.

29. Install the accelerator pump linkage. The link and pivot must be fitted on and slid over the S-link pointing outward at the top. Be sure that the link pilots on the pivot screw.

30. Check secondary throttle opening.

31. Check primary throttle stop. Adjust the primary idle screw adjustment to allow the primary throttle blades to close fully.

32. Check the fast idle linkage for proper operation. Throttle linkages are more easily checked with the carb off the engine.

33. Install the metering rods—the step-up piston, the metering rod and hanger and the step-up spring. The assembly should move up and down smoothly once assembled. Push vertically directly over each metering rod with even amounts of force when checking this operation. Remember to hold the metering rods into the arm until installed or they can fall out.
Tip 1: You may have to wiggle to get the rods to

center and drop into place. Also dimples in the arm go toward the front.

Tip 2: There is one dimple in the flat shaft to the left and right of center. The dent side should be rearward and the point of the dimple should be toward the primaries.

34. Install the center metering rod retainer. Then install the retainers over the metering rod wells.

35. Install the choke rod and clips.

36. Install the choke countershaft lever/rod into choke blade bracket. Install bracket on shaft and attach the rod and secure with screw. Tighten securely.

37. Test air door tension. Tap rear of air door several times with your finger. With some practice you can judge the spring tension by tapping to double-check your setting.

Secondary Air Valve Adjustment

38. Loosen the larger, outer lock screw. Note small screw inside large outer screw. Without special tool, it is almost impossible to release the outer lock-nut/screw and hold the current tension on spring, which is controlled by the inner screw/shaft. Use large-blade screwdriver and loosen the outer screw, the locking system. Allow the air valve to drop down—loosen and it should hang in the vertical, wide open position. No tension on the spring.

39. Turn the smaller, inner screw and its tension spring putting tension on the air door until it is just closed and then go 1 1/2" additional turns. This procedure is also given as 2 turns total measured from the vertical position.

40. Hold this position with one small screwdriver and tighten the lock screw with a second small screwdriver. Compare current tension on air door by tapping again. Secure outer lock screw using large-blade screwdriver.

41. Slide the fuel inlet screen, if being used, into the fuel line fitting and install the fitting. Tighten securely.

THERMO-QUAD DISASSEMBLY

1. Top view of the '72-and-newer Thermo-Quad. Note primaries are toward the top, choke is open and the throttle linkage is on the left. The arrow indicates the two pipes that identify it as a '72 or newer model.

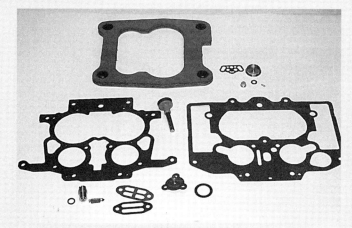

2. The gaskets and service parts for the Thermo-Quad are quite unique. There are two body gaskets to seal the 3-piece body. Large gasket at the top left is the base gasket.

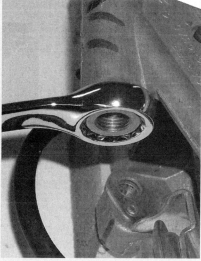

3. Be sure that the carburetor is securely clamped before using a large wrench to remove the fuel inlet fitting.

4. There is a gasket that seals the fitting to the housing that generally comes off with the fitting as shown here. Check the fitting for a filter screen and also inside the housing.

5. On top of the secondary air door, you will find a linkage that connects to the choke—called the choke countershaft fast idle lever. Remove the attaching screw while holding the linkage securely. Note the orientation of the ends of the link.

Carb Rebuilding

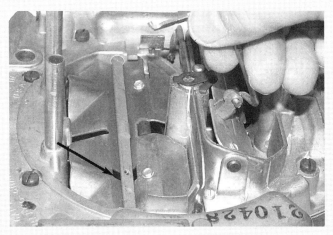

6. Remove the bracket from the countershaft and lift the link up and turn so it can be removed from the choke plate bracket. Arrow points to attaching screw hole.

7. Remove the clip from the upper end of the choke diaphragm connector rod using needle-nose pliers.

8. Remove the clip from the lower end of the rod and remove the rod.

9. Remove the screws from the metering rod cover plates, one per side.

10. Remove the screw from the retaining bracket in the center of the metering rod arm.

11. Lift the metering rod arm and metering rods straight up. Hold the metering rods in place on each end during this process.

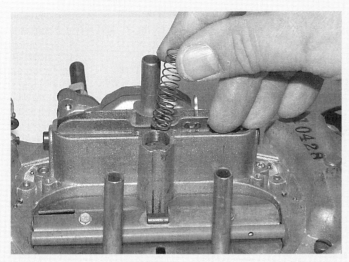

12. Lift out the step-up piston spring that stays in the well when the metering rod and arm assembly is removed.

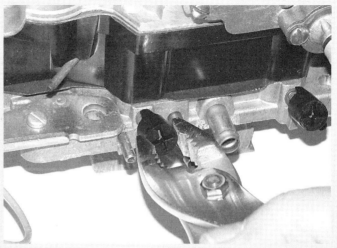

13. Using a pair of pliers, pull off the black plastic, idle mixture screw covers—one per side on front of carburetor housing.

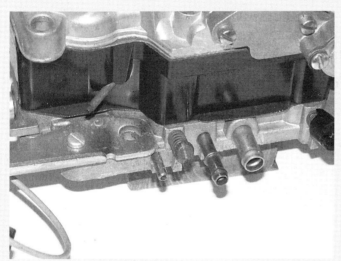

14. With the plastic cap removed, the adjustable screw head is now visible and accessible.

15. Remove the screw that holds the accelerator pump nozzles in the primary. It is easily accessible with choke plate in vertical position.

16. Nozzle assembly consists of the nozzle, the gasket and the attaching screw.

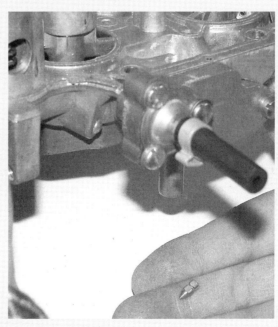

17. Once the nozzle has been removed, turn the carburetor upside down and catch the tiny check valve that should fall out.

Carb Rebuilding

18. Remove the clip from the lower end of the accelerator pump link and pull link from linkage bracket. Rotate link so the upper end of link can be removed from pump arm.

19. Remove the shouldered screw pivot and spring (if used) from pump arm and slide the arm off the S-link.

20. Leave the S-link in the pump-shaft's upper end until you are ready to remove the actual pump.

21. Disconnect the choke diaphragm hose from the fitting in the base plate.

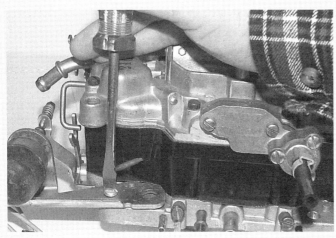

22. Remove the screw from the bracket that holds the choke diaphragm to the base plate/flange. Set choke diaphragm aside.

23. Before you take the carb apart, check the alignment of the bowl vent valve. It is located on the right front corner of the Thermo-Quad. The upper arm, shaped like a C, should be in the fork of the Y-link that comes up from the base. This relationship must be maintained at reassembly.

24. Remove the 10 screws that hold air horn to the base plate or throttle body.

25. Remember that two of the 10 screws are located between the choke plate and the wall the separates the primary and secondary sides of the carburetor.

26. The 10 screws used on the Thermo-Quad are all the same length.

27. Note that the transfer rod still connects the top to the bottom, on the left side of the carburetor.

28. Lift the top or air horn straight up off the black plastic main body. The gasket will probably stick/stay with the air horn.

29. Rotate the top around the transfer rod until the top is almost upside down.

Carb Rebuilding

30. Once the air horn is at least vertical, the transfer rod can slip out of the lever and the top and main-body/throttle body are separated.

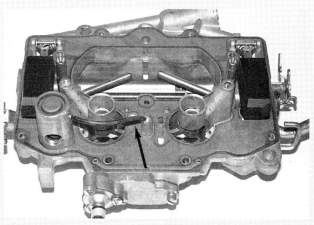

31. Turn the air horn (top) fully upside down and set in a safe place on the workbench. Note the primaries are toward the bottom. Also note the accelerator pump tube on the lower left, running to the lower center (arrow).

32. The air horn now has the primaries toward the top. Note the accelerator pump tube runs from the top center of the photo to the top right (arrow). Also note that there are two floats—one on each side.

33. The gasket stays with the air horn or top. There are 2 floats, 2 secondary jets and 2 nozzle bars that are part of the air horn.

34. Using a thin-blade knife, loosen the gasket from the aluminum housing around the full top. Be very careful. Do not try to remove the gasket yet.

35. Remove the plastic tube from the accelerator pump housing. It must be reused, so be very careful. Slip off both ends and put in a safe place.

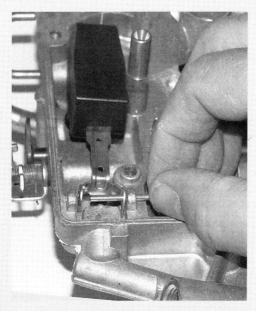

36. Remove the pin from the float and remove the float.

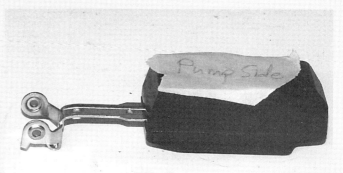

37. Mark the left or accelerator pump-side float.

38. Remove the needle, then remove the seat with a wide-blade screwdriver.

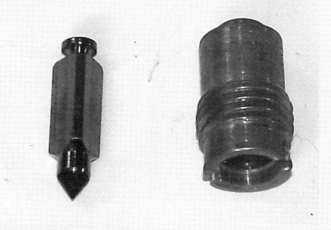

39. Keep each needle and seat together and mark so they go back into the same location if they are being reused.

40. Using a small open-end wrench, remove the secondary jets—one on each side.

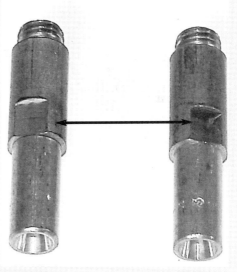

41. The secondary jet is very tall. Double-check that both jets are the same size. Numbers are etched into side of jet. These flat sides are for a wrench.

Carb Rebuilding

42. Remove the S-link from the top of the accelerator pump shaft. Flip the air horn over to its normal orientation (right side up). Support the air horn on two blocks of wood favoring the primary side. Using a hammer and flat-end punch, tap the accelerator shaft end lightly.

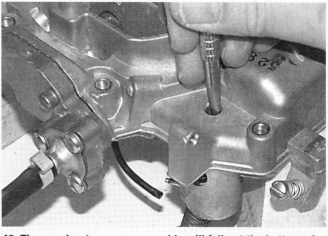

43. The accelerator pump assembly will fall out the bottom of the pump well/housing, so be ready to catch—do not allow to fall on floor.

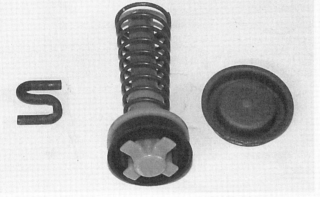

44. The 3 main pieces of the accelerator pump assembly—the S-link on the left, the pump and spring assembly in the center and the bottom seal on the right. Note that the spring is as long as the shaft so at reassembly, the shaft must be pushed up through the housing, compressing this spring and then held together by slipping the S-link into place.

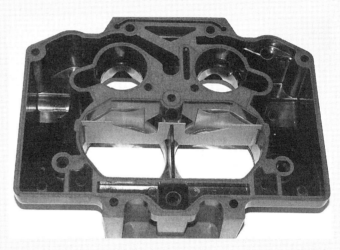

45. Lift the main body, the black plastic part, off the bottom or throttle body section and carefully set on workbench. Note the diffuser blade in the secondary.

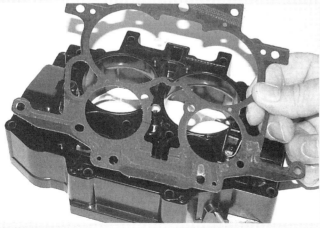

46. Carefully remove the bottom gasket from the plastic main body. This gasket usually sticks to the main body but may stay with the throttle body.

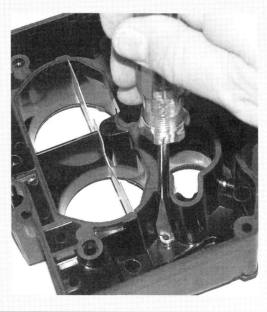

47. Remove the primary metering jets—one on each side between the primary and secondary bores at bottom of float bowl. Use wide-blade screwdriver.

Rebuild & Powertune Carter/Edelbrock Carburetors

48. The Thermo-Quad primary jet is similar to the AFB/AVS jets but not actually the same. Check for numbers stamped/etched on the top surface. Also check the jet's orifice size using number drills.

49. The secondary diffuser sits across the secondary bores. It can be removed if desired but it can also be left in place as long as it is not damaged.

50. The throttle body section (bottom) of the carburetor mainly holds the actual throttles. It is made of aluminum.

51. Remove the step-up lever and its pivot pin. It is located between the primary throttle bores in the front of the carburetor (top in this photo)

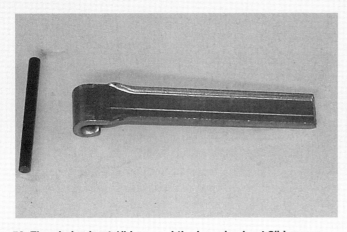

52. The pin is about 1" long and the lever is about 2" long.

53. Remove the idle mixture screws—one on each side.

Carb Rebuilding

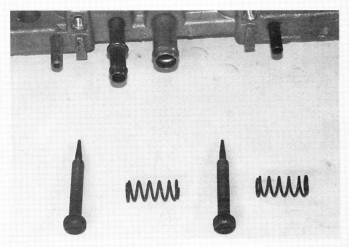

54. There is a small spring that fits under the head of each idle mixture screw.

55. Note the step-up lever cam located between the two primary throttle bores—center and toward top of photo.

56. The step-up lever cam, center, toward right, acts as part of the primary throttle and moves as the throttle is opened. The cam is offset so that it moves (lifts) the lever as the primary throttle is opened.

THERMO-QUAD REASSEMBLY

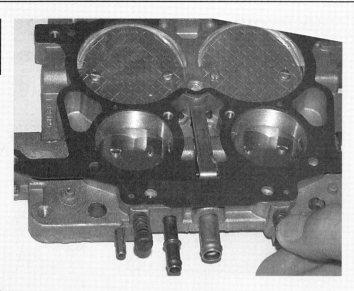

1. Test fit the new gasket to the throttle body section. Reinstall the step-up lever and pin.

2. Install the main jets and tighten securely.

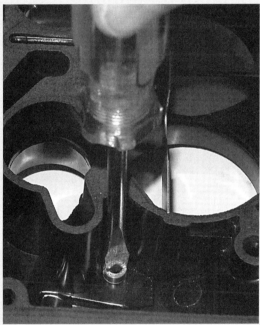

3. Install the secondary diffuser, if it was removed. Be sure that it is all the way down in its slots.

4. Test fit the main body to throttle body gasket on the bottom of the main body (plastic part). The locating dowels are on the bottom surface of the main body so push the gasket down over the alignment dowels.

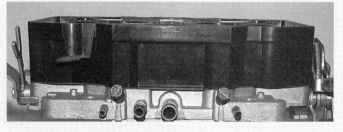

5. Install the main body and gasket onto the throttle body section and push the alignment dowels into location. Main body should sit square with the throttle body.

CARB REBUILDING

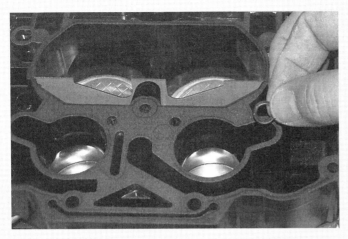

6. On the top surface of the main body, there are 2 O-ring seals that fit in the wells toward the outside of each primary throttle bore. Place these two O-rings in location.

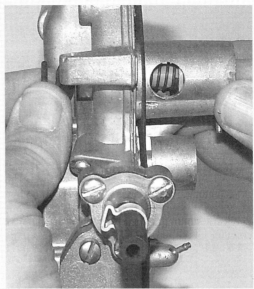

7. Install the accelerator pump and spring into the pump well. Push the pump shaft up through the top of the housing and insert the S-link to hold in place.

8. With the S-link holding the pump up in the pump well, position the pump well over the check seat (lower right).

9. Set the pump down squarely onto the check seat.

10. Tap the check seat into plate at the bottom of the pump well. Be sure that the check seat is sitting squarely against the housing all the way around (arrow).

11. Install the seat into the air horn and tighten securely.

12. Install the needle into the seat.

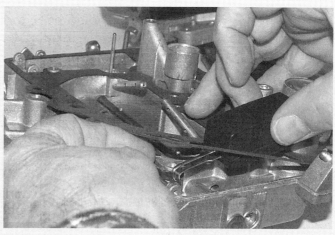

13. Install the air horn to main body gasket. It is a very tight fit so take your time, use great care and push down carefully around all tight spots.

14. Install the float and float pin. Be sure that the pump-side float is installed on the pump side (marked at removal). With the gasket installed and the air horn upside down, measure the float height.

15. If the float height has to be adjusted, bend the float arm at arrow. Do not push down against the seat.

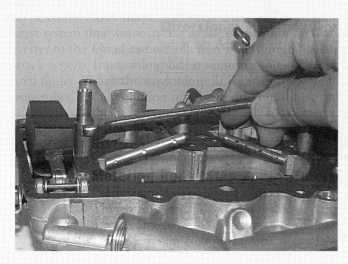

16. Install the tall secondary jets and tighten securely.

17. Install the small plastic tube to the accelerator pump cylinder. Push both ends on securely.

CARB REBUILDING

18. Be sure to attach both ends—no kinks in the tube.

18. Install the curb-idle speed solenoid located on the right-front of the carburetor.

19. Install the choke diaphragm to the base plate/flange. Note: This diaphragm and the above solenoid could be installed later but they are more easily done with the air horn off of the carburetor.

20. Double-check that the main body has the two O-rings in the proper location.

21. Insert the transfer rod into the throttle linkage slot on the left side of the carburetor.

22. Rotate the air horn over, while properly aligned, and connect the transfer link to the bracket on the air horn.

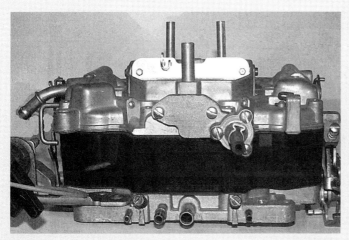
23. Rotate the air horn back so that it is level/square with the main body and lower the air horn straight down onto the main body. It should fit flush all the way around.

24. Be sure that the bowl vent arm from the air horn (top) is in the Y-slot of the arm on the throttle body (bottom).

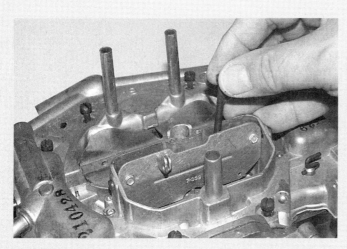
25. Install the 10 attaching screws. Do not forget the 2 screws located inside the choke bore.

26. Tighten the 10 attaching screws. Do not forget the 2 screws inside the choke well. After all 10 are secure, let the assembly sit and then retighten all 10 screws again!

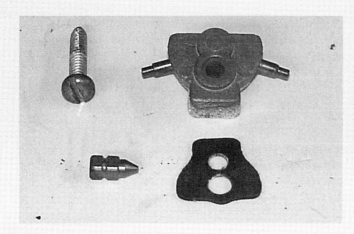
27. Install the accelerator pump check valve (lower left). The point goes down. Then install the nozzle and gasket.

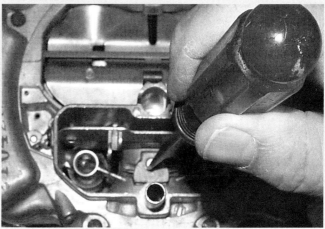
28. Tighten the nozzle screw securely. Note: Nozzles point at center of primary throttle bores.

CARB REBUILDING

29. Install the accelerator pump arm. Slide the arm onto the S-link. Be sure that the S-link is installed properly—pointing outward at the top.

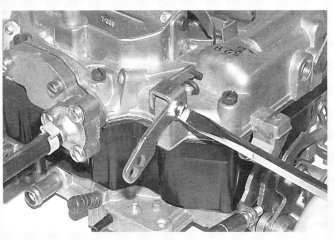

30. Install the pivot screw and pilot the arm on the screw's outer diameter.

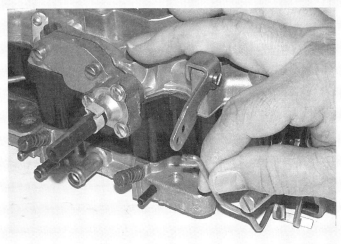

31. Install the accelerator pump link into the lower part of the linkage and then slip it into the proper hole in the pump arm. Add the retaining clip.

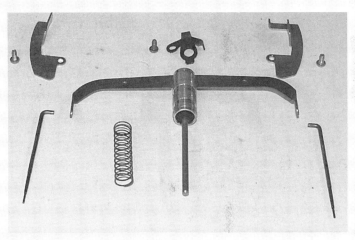

32. Gather the metering rod parts—spring, piston and arm assembly, 2 metering rods, 2 cover plates and screws, and the retaining bracket and screw.

33. Install the metering rod assembly. Remember to hold the two metering rods into the arm as you lower the assembly into place. You may have to wiggle the assembly to get everything to line up properly.

34. After the assembly is in place, push the center piston down to be sure everything is aligned properly.

35. Hold the metering rod assembly down while you install the retaining bracket and screw. Tighten screw (arrow). Note that the center metering rod adjusting screw can be accessed through a hole in the retaining bracket.

36. Install the cover plates. Note that the cover on the left side (throttle side)—top right arrow—has a tab that protrudes over the secondary air door and that the edge of the air door on this side has a slot in it (bottom left arrow).

37. Install the choke lever link into the choke blade bracket. Install the shaft bracket onto the link and then attach the bracket to the countershaft with the screw. Tighten securely.

38. Once the choke linkage is in place, hold the air door fully open using your finger and be sure that there is clearance between the top part of the air door and the choke linkage bracket (arrow).

39. The air door adjustment is on the left side of the carburetor. Use a very large blade screwdriver to loosen the large outer locking screw on the air door tension spring. Without the special tool, the Thermo-Quad air door adjustment is much more difficult than the AVS.

40. Once the outer lock screw is loosened, the inner screw can be adjusted using a smaller screwdriver. I would recommend placing a piece of tape on one side of the screwdriver blade to make it easier to count full revolutions/turns. Tip: The basic spring setting is 1 1/2 to 2 turns from hanging vertical. Tip: Once you have actually set the desired spring tension using the small screwdriver and the inner spring, you can use a really small screwdriver to tap the outer slot of the outer locking screw enough to hold the inner screw in place. Then the small screwdriver can be removed and the large screwdriver inserted to lock everything in place securely.

41. The amount that the air door opens can be checked using a steel scale and measuring from the tip of the air door blade to the rear-side of the choke well holding the air door fully open with your finger.

42. Before you install the carburetor onto the engine, it is a good idea to check the throttle linkage on the left side of the carburetor. Check to be sure that the throttles fully close and that the idle stop isn't holding the primaries partially open. Then check the basic operation of the fast idle linkage. This is much easier to do before the carb is installed on the manifold.

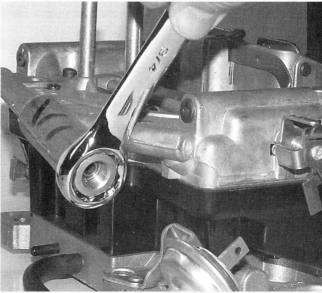

43. Install the screen, if being used, and the fuel inlet fitting and tighten securely. Be sure to clamp the carburetor securely before using wrench.

Chapter 5
Carb Hardware & Installation

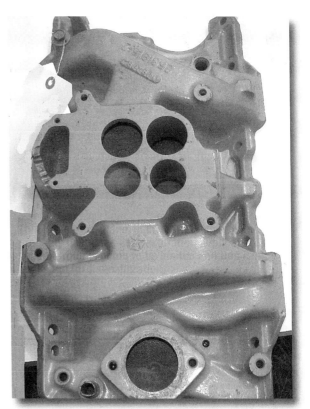

Production intake manifolds tend to be made out of cast iron. This is a late-1960s 340 Dodge/Plymouth design for an AVS carburetor.

The carburetor sits on top of the engine's intake manifold, and the manifold is an important part of the induction team. The intake manifold bolts to the engine's cylinder heads but there are other parts that are essential to the basic installation. Air must be delivered to the carburetor through an air cleaner and hood scoop system. Fuel must be delivered to the carburetor from a fuel supply system consisting of filter, fuel pumps, fuel tank and fuel lines. While both of these supply systems are related to the vehicle that the engine is installed in, they are both very important to the proper operation of the carburetor.

INTAKE MANIFOLDS

There are so many intake manifolds to choose from. There are two-, four-, six- and even eight-barrel manifolds available. Since we are dealing with a four-barrel carburetor, I will limit the discussion to four-barrel and eight-barrel intakes. In general these intake manifolds fall into two main styles—dual-plane and single-plane.

Dual-Plane

A dual-plane manifold by design has the intake runners laid out in two levels, an upper and a lower. For the typical V8 engine, this puts the runners of 4 cylinders on each level. The left side of the carburetor (one primary and one secondary) feeds one level and the right side feeds to other level. Typically they are tuned for every other cylinder in the firing order grouped into one level. The single-plane is much easier to explain and visualize since all runners come into the central, common plenum.

Production Dual-Plane—Most production four-barrel intake manifolds are dual-plane designs and are made of cast iron, with a few exceptions. While high-rise name became commonplace with the later introduction of aluminum aftermarket manifolds, many of the late 1960s and newer cast-iron production manifolds were high-rise designs. "High-rise" refers to the height of the carburetor pad above the cylinder head inlet ports. The first high-rise intake manifolds were introduced on small-block engines because they had more hood clearance. High-rise manifolds raise the carburetor height closer to the hood.

The second aspect of production dual-plane intake manifolds is the carb pad itself. Most production manifolds have four individual holes in the carburetor pad. There are two oval hole versions, typically divided left and right, rather than round holes. There is also the standard four-hole pattern that features four holes of about the same size and the Thermo-Quad pattern spread-bore that uses two very large secondary holes and two small primary holes.

Aftermarket Dual-Plane—Aftermarket intake manifolds are usually made of aluminum. They can be either dual-plane or single-plane. The majority of the performance dual-plane intakes made in the last 30 years have been high-rise designs. The aftermarket rarely uses the four-hole style of

CARB HARDWARE & INSTALLATION

Most production intake manifolds feature a dual-plane design. Note that the carburetor pad has two openings, one to the left and one to the right. The left side feeds the upper level of the dual-plane. Front of the manifold is toward the bottom.

This is also a cast iron, production dual-plane manifold, but notice the hex-headed fittings in the bottom of the plenums. The one on the left appears halfway between the primary and secondary bores while the one on the right appears below the primary bore. These are the EGR (exhaust gas recirculation) jets. They tap into the heat crossover passage which runs left to right in the middle of the manifold.

This is also a production-style dual-plane intake manifold but it is designed for the Thermo-Quad. Note the very large secondary throttle bore holes, toward the top, and the much smaller primary throttle bore holes toward the bottom.

This is a typical aftermarket aluminum dual-plane four-barrel intake manifold. The front is toward the top. Two things to note. One, it is designed for a 426 Hemi or individual runner heads rather than the joined runners common on wedge-style intakes. Two, how the runners are joined is more obvious—the upper plenum feeds to two end cylinders on the left (left front and left rear) and the two middle cylinders on the right side.

carb pad, opting for a two-hole design with the divider running front-to-rear. There are still the two basic shapes—standard and spread-bore like the Thermo-Quad, which has small primary openings and very large secondary openings.

Racing Dual-Plane—This is the newest member of the dual-plane family, introduced in the late '90s.

You could consider it a taller high-rise style. The main difference between a racing dual-plane and a typical performance aftermarket dual-plane is the height and runners that tend to be larger in cross-

An aftermarket aluminum single four-barrel dual-plane designed for '61–'62 348–409 Chevy engines, the so-called W-series engines.

This is a wedge dual-plane intake but sits higher above the block than the standard high-rise dual-plane. This style of manifold is considered a race dual-plane but they can be used in street and strip applications.

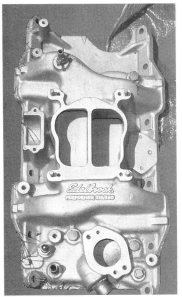

The typical aluminum aftermarket four-barrel dual-plane intake designed for wedge heads joins the first two cylinders on each side and the last two cylinders on each side. Front is toward the bottom. The pairing for the upper level (left plenum) is still the two end cylinders on the left side and the two middle cylinders on the right.

sectional area. I have used the race terminology only to give the group a label. While they can be raced, they tend to be more streetable than single-planes.

Single-Plane

Production Single-Plane—There are only a few production manifolds that were built as single-planes. Generally these date from the early to mid-1960s. Most were made of cast iron. When placed side-by-side, these manifolds would not be considered high-rise, so I'll list them as low-rise or standard-rise designs. Unless you are doing a resto project, I would not use any of these manifolds.

Aftermarket Single-Plane—These are perhaps the most common aftermarket intake manifold. They are made of aluminum. The carb pad is almost always a one-hole design—basic square with rounded corners.

There are single-plane four-barrel intake manifolds available for almost every V8 engine built in quantity. They are available from Edelbrock, Weiand, Offenhauser, Holley and the performance parts divisions of the OEMs like Chevy, Ford and Mopar (Chrysler). While each specific design and shape has its advantages, most engines will utilize one that works best for power and another that works best for torque (street). This type of recommendation is unique to each engine and can usually be obtained from an engine book on the specific engine desired. This recommendation would come from testing on a dynamometer or in-

Another aluminum dual-plane. This one is designed for individual runners (not a wedge style) and fits big-block Ford engines. Notice that the runners themselves are not attached to the bottom plate underneath the runners and air can get in at the front (left center) and rear (right center) and between the runners themselves. This style of manifold is called an Air Gap design by Edelbrock.

This is an aftermarket single-plane intake manifold for wedge cylinder heads and a Thermo-Quad carb—front is to the left. All runners come into the common plenum.

Carb Hardware & Installation

This is an early-style dual four-barrel or eight-barrel single-plane intake manifold. It is for wedge cylinder heads.

Another wedge single-plane but designed for the standard AFB/AVS carburetor flange and throttle bore sizes.

This is also an early dual four-barrel inline intake manifold designed for wedge heads (383–413 style) with the two AFB production carburetors mounted. Note that the choke is on the right rear of the carburetor—the blade is closed on top of carb and choke unit at top right.

This is considered a race single-plane intake design for wedge cylinder heads. While race single-plane tend to be taller than standard ones, that aspect is hard to show in a photo. Race single-planes also tend to have taller/bigger runners and the designers tend to extend the runners further into the central plenum as shown here. The runner dividers are almost touching at the bottom of the common plenum (center).

made of cast iron. While impressive visually, I would not recommend using one of these today unless you are doing a restoration project.

car or at the race track.

Racing Single-Plane—Racing single-plane manifolds are much taller than the standard street designs and have much larger runners. Because of the basic height of the manifold, these intakes may cause hood-clearance problems and may require a hood scoop. Many of these race intake manifolds are designed to work with a specific race cylinder head, port size or design.

Eight-Barrel or Dual Four-Barrel Single-Plane—The eight-barrel intake system can come in several variations, but the dual four-barrel inline system is probably the most popular. They were somewhat common in production (limited options) in the late 1950s and early 1960s. Most of these intakes were based on the single-plane style and

Dual-Plane Eight or Dual Four-Barrel

These are perhaps the newest group of intake designs. They have only been available since around 2007. They feature the high-rise style of high-performance manifold and also have a dual-plane feature that makes them much better suited to street and dual-purpose applications. While the high-rise design makes them taller than the originals, they are not as tall as the race single-plane discussed earlier and should not cause problems with hood clearance in most vehicles, but check it out just to be sure.

Cross-Ram—The only known production versions of the cross-ram dual four-barrel system were the 413 and 426 Max Wedges in 1962–'64. There were race versions used on the 426 Hemi and the 390 AMX for the Super Stock classes. Edelbrock once produced a line of cross-ram manifolds called the STR series, but they are hard to find today.

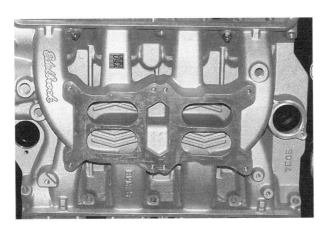

This is a dual-plane eight-barrel inline manifold for an individual runner cylinder head—big-block Ford. Front is to the left. It is not easy to see that the upper plenum is the lower halves of the two carb-pads and joins the 2 end cylinders from the upper row and the two middle cylinders on the lower row.

This is an eight-barrel cross-ram package as built for the '62–'64 413–426 Dodge Plymouth Max Wedge cars. The bell-crank in the middle opens both AFB carburetors at the same time.

This is another eight-barrel inline dual-plane intake (for wedge heads) and it is easier to see the two level—upper and lower. The two end cylinders on the bottom row are connecting to the upper plenum.

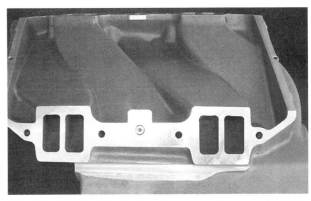

The 413–426 Max Wedge heads use very large ports. This is a cross-ram manifold that is designed by A and A Transmission that fits the standard sized 440 heads. Notice the extra material around the ports that allows for the port-core change and the basic manifold can be cast with the larger ports also.

Intake Manifold Prep

Typically you would buy your intake manifold as a finished piece that has been cast and machined rather than welded from sheet metal. Therefore, you might think that there is nothing to be done to the manifold to prep it for installation onto the engine. Since the manifold has more to do than just hold the carburetor on top of the engine, there are potential issues. In most engines, the intake manifold also serves as the cover for the tappet chamber. Since there is all kinds of oil spraying around the tappet area, sealing the tappet chamber is an issue. This is made more difficult by the fact that very few performance engines are built without being milled, which can greatly change the sealing geometry. Ideally, the ports in the manifold should line up with the ports in the heads. Since many cylinder heads today are ported, this may require the intake ports to also be ported or just matched to the heads or it may require machining the port faces of the manifold.

Milling—Almost all performance engines have milled parts. The intake manifold fit is affected by any block deck milling and any cylinder head milling. Generally these two amounts (the amount milled off the block's deck surface and the amount that you machined off the cylinder head's deck face) should be added together to come up with a total amount milled off the mating surfaces. Most engines have a formula for how much to mill off the intake manifold face of the cylinder head to fix the basic fit issues. This manifold face milling is not always done. Additionally the front and rear tappet valley walls (front and rear) on the block should be milled (per formula). Another potential problem with used parts or even rebuilds is that you may not know if the parts have been milled prior to your purchase and use.

Generally, the cylinder heads and block are milled. When the engine is preassembled with gaskets, the intake manifold must now fit with the

Carb Hardware & Installation

MANIFOLD TUNING

What engine builders are trying to do with the intake manifold design is tune it for more power or a better power curve. First you must sort out the fuel distribution. There can be much bigger gains in good fuel distribution than in dimensions. Then you have the runner's length, cross-sectional area, taper and angle (if any) along with the plenum's shape, size and interconnection, if this applies. As a general consideration, changing the intake manifold's runner length by 3/4" changes the peak tuning number (rpm) by 500 rpm. Longer runners make more torque (lower rpm) or shorter runners make more the power and the peak numbers occur at higher rpm. In general, it is probably of more value to port match at the manifold face than it is to change the runner length.

The breastplate is a stamped-steel sheet that serves as an intake gasket on big-block Chrysler engines—383s and 440s. The section that goes between the cylinder head and the intake manifold serves as the intake gasket, but the bottom section serves a second purpose of keeping the hot oil off the intake runners and also allows air to circulate under the runners.

This intake shield goes on the bottom of the production intake manifolds used on the 426 Hemi engine. It is designed to keep the hot oil off the bottom of the intake runners.

other engine pieces. As a general rule, you do not want to machine the intake manifold to fit the specific parts being used. Any machining on the intake to make it fit makes it unique to this specific engine. Additionally any manifold change would require that the new or test manifold would have to have the same machining done to allow it to fit before you can do any testing. If the intake manifold doesn't fit properly, that usually means that something is wrong—a more common problem with used parts (blocks and heads) than new ones. Figuring out the WHY is not easy. If the head has been machined too much or the block decked too far, you can't add material back onto these surfaces, so machining the manifold or utilizing extra think intake gaskets may be your only solution.

The first concern in intake manifold fit is if all the intake manifold attaching bolts fit properly into the cylinder head holes that will allow you to tighten the bolts. This has to be your first concern. The second concern is to decide if the ports in the intake manifold line up with the ports in the head. Port-matching made the ports the same size and shape but they may not line up—vertically. This is much more easily seen by looking down a runner on a single-plane manifold. On most dual-plane intakes, you hunt for the cylinder that allows you to see the intake manifold/cylinder head interface. Since you have made the ports the same size and shape, the port may need to move up or down. The key to this alignment is the intake gasket. If it needs to move up, then use a thicker gasket. To move it down, you could use a thinner gasket but that may not be available, so you might have to remachine—see the Manifold Fit sidebar on page 104.

Breastplate—Some engines use a breastplate style of intake manifold gasket like the 383 and 440 Chrysler/Mopar big-block engines. This breastplate is generally a steel stamping. This style of intake gasket serves two main purposes—to seal the intake manifold to the head and seal the tappet-valley opening which is the main function of an intake gasket—to seal. The second purpose is to keep the hot oil that's flying around the tappet area from hitting the bottom of the intake runners and heating them up.

Heat Shield—On some intake manifolds, there is a heat shield that is added to the bottom. In recent times, it has been an addition by racers but it actually dates from the mid-1960s on the 426 Street Hemi in production. It serves the same purpose as the breastplate discussed above in that it keeps hot oil off the bottom of the intake runners.

Air Gap—Edelbrock uses the term Air Gap to describe a new series of intake manifolds. However, an actual air gap on the intake manifold actually dates back but to the 1970s, perhaps even to the 1950s. To simplify the explanation, an air gap manifold casts the breastplate to the actual intake manifold as a one-piece unit. That means that air

MANIFOLD FIT

With most wedge-type V8 engines, manifold fit is less of an issue, but with engines that use vertical intake manifold attaching screws, like the 426 Hemi, the manifold fit leads directly to engine performance, both power and torque. For as much time as you spend working on the cylinder heads, you would think that you should spend a similar amount of time getting the intake manifold ports to line up with the ports in the head. Usually by this time, you only worry with how the manifold fits as a function of getting the screws started. Having the bolts line up does not guarantee that the ports line up. The holes for the bolts tend to be larger than the bolts themselves, so there is some float. So let's assume that we make the holes much bigger. Now you set the manifold on the gaskets, and look down the ports to see how they line up. If they don't, what do you do? If the port in the manifold is high or above the one in the head, then we can mill the manifold face of the intake manifold. But what if it is too low, already below the port in the head. We can remove the head and mill it but that doesn't sound like a first choice. A thicker intake gasket would do the job—such as 0.060", 0.090" and 0.120—but unfortunately they are not always available for all engines.

How To Measure—Manifold fit is more critical on a 426 Hemi than any other engine. With the manifold supported by shims on the front and rear tappet valley walls of the block, the gaps can be measured and indicate what needs to be done. Obviously the gap at the top of the manifold face should be equal to the gap at the bottom on all four corners. This same technique can also be used for all wedge engines.

Tip: The 426 Hemi intake manifolds have always been hard to seal. If they leak, then the engine draws oil into the combustion chamber, which acts like a low octane fuel and leads to detonation. Adjusting the carburetor calibration can't fit this oiling problem. The little 1/4" screws that attach the manifold must be torqued in sequence and in inch-pounds. But that is not all there is to it! I would recommend spending a little extra time fitting the manifold. This technique works with either the wedge or Hemi engines. Most of the manifold fabricators use a large fixture to make sure that the finished product is square and has flat sealing surfaces. Since the manifolds do not like heat in general, and don't like to get banged around, it is a good idea to check the assembled alignment of your manifold on your engine.

Set the manifold on your block and heads (using a used head gasket), and space it up off the front and rear tappet valley walls about 0.040". Then center the manifold—front to rear and left to right. Then measure the gap at the top and bottom of the intake sealing surface or gap (where the intake gasket would normally go) on the four corners (left front, left rear, right front and right rear). Don't forget to check each location top and bottom to make sure that the milled angles are correct or at least matching. (That's eight measurements.) Since the typical intake gasket is about 0.060""thick, the variations can probably be a few thousandths, but you might want to tweak it if it is more than maybe 0.005". Also, don't forget the two end seals with little foam gaskets. Thicker intake gaskets are available from several manufacturers.

Note: Each engine and cylinder head combination has unique aspects to its intake gasket layout but most use four separate gaskets—one on each side and one at the front and one at the rear. Sometimes the OEM factory performance parts group (GM, Ford or Mopar) offers a wider selection of gaskets for their engines.

Tip: Almost every used 426 Hemi block, cylinder head and intake manifold has been milled and it can become very difficult to figure out the amount. This can be a common problem for any engine that a high percentage of production has been used in racing or performance activities. If the head and intake manifold surfaces are parallel, then the intake gasket has a pretty good chance of sealing it. The weak link can be in the four corners where the intake surface, the deck surface and the block's tappet valley walls all meet. A small amount of RTV in these four corners can help seal. Wipe away any excess because it might get into the oiling system.

can pass below the runners from front to rear with the manifold installed on the engine. Now the bottom of the air gap manifold shields the runners from the hot oil in the tappet valley area.

Manifold Modifications

While you do not have to do much to the manifold for preparations other than the basic fit considerations discussed in the box above, there are a couple aspects that you might want to consider before you install the manifold.

EGR Jets—On some production intake manifolds built from 1972 and newer, there may be an EGR (exhaust gas recirculation) jet or two in the bottom of the plenum. Because of the emissions rules these tended to only be in spread-bore designs. They generally only occur in cast-iron production manifolds rather than aftermarket aluminum versions. If they are not going to be used, they should be removed and plugged. Remember that it is an opening to the outside so it acts just like a vacuum leak if not sealed properly.

Carb Hardware & Installation

This is an Air Gap Edelbrock manifold that allows air to circulate under the runners from the front (toward left) and rear (toward right) and between the individual runners on each side as well.

The EGR (exhaust gas recirculation) jets are the two hex-shaped objects in the bottom of the two plenums—one on each side. They tap into the heat crossover which runs left to right across the center of the manifold in production intake designs.

Divider—Generally, an aluminum, dual-plane intake manifold will have a divider between the left and right sides of the manifold. Single-plane manifolds generally do not have dividers—one common square hole. Cutting the divider down, but not all the way, will generally make more power without affecting the torque. The divider runs from the front to the rear of the manifold and separates the left and right side of the manifold. You don't want to cut the divider all the way back to the rear or front wall (leave points on each end) and you don't want to cut the wall down to the top of the lower plenum (leave 3/4" high wall).

Distribution Tabs—There is going to be some confusion between fuel distribution tabs and fuel distribution dams. Tabs are added to the carburetor while dams are added to the manifold plenum. Only production carburetors have fuel distribution tabs. They are added to the outer surface of the venturi. They can be very hard to see and you must

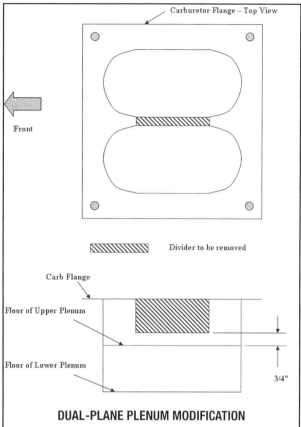

DUAL-PLANE PLENUM MODIFICATION

To make more horsepower, you generally want to cut down the center divider on a high-rise dual-plane intake manifold. However, you do not want to cut it completely out. It is cut out as more of a large notch—as shown in this drawing. Shaded area indicates material to be cut out.

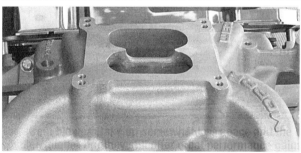

This race dual-plane has the divider cut out.

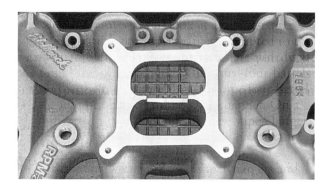

This Air Gap manifold also has the divider cut out or cut down. It ends up to be a big notch, not completely removed.

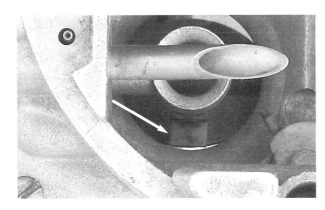

The fuel distribution tab is added to the carburetor's venturi as shown. Only production carbs use these tabs.

Fuel distribution dams are usually made of wood—small, thin and cut to length. They are usually made from popsicle sticks which gives them their common nickname. This open plenum shows two fuel distribution dams—one at the top right and one at the top left. One or two dams are the most common while no dams is very rare. The number, size and location must be determined with extensive dyno testing. No two engine designs are the same.

Usually only one distribution tab is added but in some cases there can be more, such as this example with one on the primary and secondary venturi.

look straight down the throttle bores from the top to be able to see them.

Distribution Dams—Generally fuel distribution dams are added to the open-plenum, single-plane manifolds. They are added to the floor of the plenum. The information on how and/or where to add fuel distribution tabs must come from an engine dynamometer. The dyno records the exhaust gas temperatures at various rpm under full load and the distribution tabs are designed to make the fuel distribution within the manifold equal in each cylinder. This is very unique to a specific engine and manifold combination. You cannot guess at fuel distribution dams—they must be tested—before and after. Don't try to copy dams that you've seen in another manifold. For example, a small-block or short-deck engine may have two dams added while a big-block or tall-deck version may not require any dams, so adding them would make that engine worse. Each dam must be tested on the actual engine with the specific carburetor that is going to be used. With the dams properly located, the engine should make more power and/or more torque. Typically they are made out of small pieces of wood—perhaps 1/4" to 1/2" high and thin, and cut to a specific length. Popsicle sticks are often used for this purpose, so the dams have been given this nickname. Once cut to size and length, they are epoxied into place. Avoid screws or pop-rivets and steel metal (steel or aluminum) in distribution dams. If a dam breaks loose, it should be able to pass through the engine without causing a failure.

Manifold Attaching Bolts—The 426 Hemi uses small 1/4" x 20 screws to attach the intake manifold. On the cross-ram, there are several lengths while the inline package uses all the same length. Most wedge manifolds use the same length attaching bolts but large manifolds like cross-rams may be exceptions. Check your engine's specific screw length by installing the screws into your manifold on the bench and see if they extend below the manifold face properly. Don't forget the 0.060" thick intake gasket.

Note: High quality bolts are available from aftermarket suppliers such as A1 Technologies, ARP and Gardner-Westcott.

Manifold Height

Intake manifolds are often called high-risers, standard or high-riser descriptions are not too helpful if you are trying to figure out your hood clearance. To figure out hood clearance, you need specific numbers and to do that you need a common baseline. The common baseline is provided by the front of the block or front tappet chamber wall. Measure from the front tappet chamber wall up to the center of the carburetor pad. You measure to the center of the pad because there could be a carburetor pad angle involved.

CARB HARDWARE & INSTALLATION

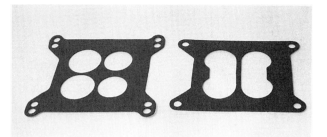

The gasket on the left would be considered a 4-hole design while the one on the right is a 2-hole design.

These two gaskets would be considered 1-hole designs, but the one on the left is for the AFB and AVS while the one on the right is for the Thermo-Quad.

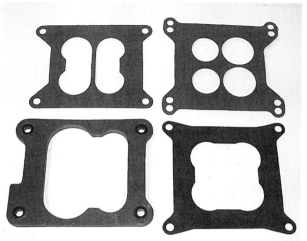

The Thermo-Quad service gasket, lower left, is actually a gasket, a spacer and an insulator. It is about 1/4" thick so it serves as a small spacer. The top and bottom surfaces have gasket material printed on and the spacer between the two gasket surfaces is made of a heat-insulating material to help keep heat away from the carburetor. The original production package used a thermoplastic spacer to keep the heat out and a separate gasket on top and bottom. The service package now replaces these three parts with a one-piece design.

Most production manifolds use a 2 to 3 degree angle while most aftermarket manifolds are zero. See page 108.

CARBURETOR SPACERS

Carburetor spacers usually match the carburetor pad used on the manifold. There could be a 4-, 2- (fairly rare) or a 1-hole spacer. There are standard and Thermo-Quad shapes and they come in various thicknesses and in two materials—aluminum and plastic (phenolic). There is also the TQ adapter, which could be considered a spacer. Spacers are available from Pro Gram Engineering, Jomar Performance or Quick Fuel Technology among others.

Style

There are two main styles of carb spacers, that go between the carb flange/bottom and the top of the manifold—a 1-hole and a 4-hole. The 1-hole spacer has one large hole for all four throttle bores. The 4-hole spacer has four individual round holes, one below each throttle bore. While they look similar, they do not tune the same. Generally carb spacers are used for fine-tuning your induction package.

Note: The spacer's effect on dual-plane and single-plane manifolds may not be the same.

Thickness

The second aspect of either style of spacer is thickness. They can come in 1/2", 1" or 2" thicknesses plus some others. The most popular one is the 1" spacer.

Caution: Consider hood clearance in addition to basic performance when selecting a spacer. Also consider the raised height on your throttle linkage since the cable now has to reach higher.

Material

The third aspect of carb spacers is material. Most are made of aluminum but many are made of phenolic plastic. It acts as a heat barrier, which is very important for a street car cruising in the summer heat. I would recommend upgrading to thermoplastic, especially for street applications. They are available from Pro-Gram Engineering and several other manufacturers.

Adapter

There are several spacers that serve the additional purpose of being an adapter. There are two-barrel adapters that allow you to use a two-barrel carburetor on a four-barrel intake and also an adapter that allows you to use the standard AFB or AVS carburetor in place of a Thermo-Quad. This TQ adapter can be used to allow a Thermo-Quad

The plastic spacer on the left helps keep heat away from the carburetor and the Thermo-Quad design on the right is slightly thicker but includes a gasket surface on top and bottom. Most performance spacers are thicker at 1/2" to 1" but there are also 2" versions.

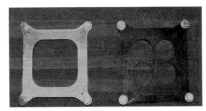
The 1-hole design spacer on the left is machined out of aluminum while the 4-hole design on the right is made of plastic.

This is the Thermo-Quad adapter for installing a Thermo-Quad carburetor to a standard AFB or AVS manifold, or an AFB or AVS carburetor to an existing Thermo-Quad manifold. It is made by Edelbrock.

This is a thermoplastic carb spacer—4-hole design and 1" thick.

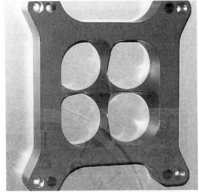
This spacer features a 4-hole design and is cnc-machined for improved airflow.

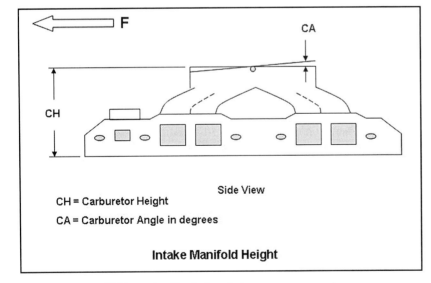
To compare manifolds and estimate hood clearance, you need a common baseline, which is the front tappet chamber wall of your engine block. Then measure up to the center of the carburetor pad. To measure the pad angle, compare the measured pad angle to the angle along the top of the intake runners on the cylinder head face.

carb on a standard manifold pad, but since Thermo-Quads are getting hard to find, I would guess that it would be more popular today to use it as an AFB/AVS to TQ manifold adapter. Edelbrock offers one.

CARBURETOR LINKAGE

Most production Carter carburetors were made to work with specific engines and models. Edelbrock's one-size-fits-all aftermarket approach requires adapters. The most common is the Chrysler linkage adapter. There is a similar one for certain Ford models. The adapter attaches to the Edelbrock carburetor and the in-car linkage attaches to the adapter.

In some cases, usually with aluminum manifolds, there can be a clearance issue between the carburetor linkage on the left side of the carb and the aluminum intake manifold casting. It is always a good idea to sit the carburetor on top of the manifold that is going to be used and check to make sure that the throttles can be opened easily and the linkage clears.

The typical four-barrel intake manifold carburetor linkage uses a bracket at the rear of the intake manifold. It is common to use two of the manifold bolts to hold this bracket in place. On the high-rise, high-performance aluminum intake manifolds, the runners are often larger and come up from the head at a steeper angle, which can cause clearance issues with this bracket in either case.

Obviously, the carburetor linkage for the two four-barrel systems is not the same as a single four-barrel. While the cable is common, the linkage that opens the carburetors is quite unique. The inline linkage system uses all the linkage on the left side with the rear carb having the choke. The production carb linkage allows the rear carb to open first and then it opens the second carb. This is a good system for the street but in racing, the racers tend to make them both open at the same time.

CARB HARDWARE & INSTALLATION

A typical 14" air cleaner designed for a single four-barrel carburetor installation. They are usually made of either stamped steel and aluminum.

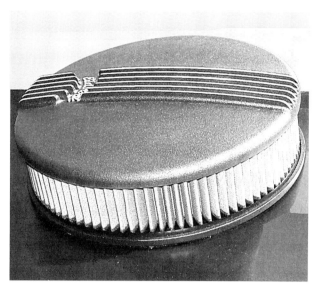

While many air cleaners are stamped-steel and then painted or chrome-plated, there are also low-restriction units that have styled, cast-aluminum tops.

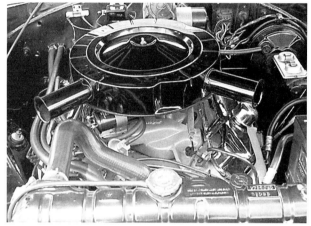

This is a production dual-snorkel air cleaner. Typical standard units used a single-snorkel design.

The cross-ram system uses a center bell-crank in the middle of the manifold and both carbs operate together (Mopar, Bouchillon Performance, Lokar Performance Products and Mancini Racing). The throttle cable pulls on the bell-crank arm and the bell-crank opens both carbs at the same time.

Throttle Springs

You would like to have two throttle springs that help close the carburetor's throttle. They are generally designed to operate one inside the other. That way if one spring fails, the other spring will still close the throttle. A throttle spring is usually anchored to a bracket that is bolted to one of the intake manifold attaching screws at the front of the manifold. You would like the two main throttle springs to be in a direct line with your throttle cable.

AIR SUPPLY SYSTEM

The key issue with the air supply system is to decide whether you plan on using a hood scoop or you will stay with underhood air. The air cleaners are different between these two systems. While most hood scoops provide cold air to the carburetor, properly designed hood scoops also provide forced air.

Air Cleaners

There are many different styles of air cleaners. Air cleaners are actually assemblies. The actual element of the air cleaner assembly doesn't change much whether it is an under-hood or fresh-air unit. However, I will consider that an air cleaner is an under-hood assembly and units that are used with a fresh air system will be discussed with hood scoops.

Almost all four-barrel carburetors have a 5 1/4" air horn—the top of the carburetor where the air cleaner base sits. I would consider a 14" x 3" air cleaner (14" diameter, 3" height) to be the baseline or standard air cleaner. The reason that the air cleaner is important to the carburetor is that it affects the calibration and it affects the airflow. There are many sources for air cleaners such as Moroso Performance Products.

Most single four-barrel, underhood air cleaners are pretty simple, 12–14" round units. The lids on the basic air cleaner are painted engine color, or black, chrome-plated, or even cast with fins or other styling features. The taller the element, the better the air cleaner is from an airflow standpoint but taller elements can cause hood clearance problems. The standard 12–14" round air cleaner is somewhat of a baseline. Air cleaners that are smaller in diameter and don't tend to flow as much air.

The production single four-barrel air cleaners tend to be more complicated and somewhat unique to each vehicle. They tend to have one or two snorkels that extend off to one side or in two directions. Passenger cars tend to have somewhat short elements because of hood clearance, while trucks tend to have taller elements (and less

This is the one-piece, 426 Hemi cross-ram air cleaner that covers both carburetors. The element goes around the outside of both carburetors, which gives it lots of volume. It may be the best air cleaner ever installed on a production engine.

These are the more common dual air cleaners used on cross-ram intake systems. This one is a '63–'64 426 Max Wedge system.

An air horn is often called a velocity stack or ideal entrance. This one is popular on the cross-ram intake systems and has a fine screen stapled cross the opening.

An ideal entrance is basically the bottom to an air cleaner assembly. This is an example of a stamped centered design.

This ideal entrance or air cleaner bottom in not on-center and could be helpful to solve clearance issues.

The econo-style of ideal entrance is made from an old production air cleaner bottom with the vertical wall around the outside cut off.

Dual-inline air cleaners pose unique problems. Most dual-inline systems from the early '60s and late '50s used two separate air cleaners. This tended to make them smaller because there is less room for two parts in the same space. The GM, Ford and Mopar wedge-inline systems seemed to follow this one air cleaner per carburetor unofficial guideline. The exception is the '66 through '71 426 Street Hemi engines. This package used one large, oval air cleaner assembly around both carburetors. It fit in a big oval shape. While most of these engines came with some form of fresh-air and hood scoop, there were many that did not come with fresh air, and therefore no hood scoop, and this created a unique underhood air cleaner. The 413–426 Max Wedge cross-ram engines and the 426 Hemi cross-ram engines tended to have hood scoops and fresh air systems, but there were options that did not include the hood scoop and fresh air, and these engines had larger, round air cleaners, which were one piece and very nice.

Tip: On some of the 426 Race Hemi engines that used the cross-ram intake manifold, they came with a one-piece, large oval intake manifold. Dyno tests have shown that this air cleaner only costs a few horsepower even at 800 hp outputs. That should qualify it is the best air cleaner ever used in production. See upper left photo.

If you use a single four-barrel package, with standard 12"–14" round air cleaner, and hood clearance is a concern, Mr. Gasket makes a dropped base plate for single four-barrels that lowers the air cleaner over the carburetor and provides extra hood clearance.

Air Horn—An air horn is also called a velocity stack or ideal entrance. An air horn or velocity stack

snorkels) because they have lots of hood clearance. Some cars that have large engines installed into small body styles or small engine compartments tend to have no room left for air cleaners, so the air cleaners are really compromised.

Carb Hardware & Installation

There have been many hood scoops used in production models. This one is on the '69 440 six-barrel cars (Dodge and Plymouth) and is one of the more popular ones.

This is called a shaker hood, because the scoop attaches directly to the engine and sticks through the hood so it can be seen to shake with the engine. These were popular on the early 1970s E-Body Chrysler products.

Perhaps the largest hood scoop used on a production package was this one from the 1968 426 Hemi Super Stock cars, Hemi Cuda and Hemi Dart.

Not all hood scoops have to be forward facing. This one picks up high-pressure air from the windshield.

usually means a part that was specifically made to act as the entrance for air into the top of the carburetor. Simply put, this item is the bottom of an air cleaner, with no top and no element. Obviously this is for drag racing only use. You are trying to help the airflow into the top of the carburetor which should help the engine increase its performance.

Note: For the econo ideal entrance, use an old air cleaner base but remember that you must hold it in place—springs perhaps—or it will fall off with hard acceleration.

Hood Scoops

With the tall carbs and manifolds used in racing, hood clearance is always a concern. Hood scoops can sometimes help this clearance situation. The hood scoop's primary function is to provide the engine with cold forced air. This is not as easy as it seems, particularly the "forced" part. Perhaps the most unique hood scoop was the shaker style hood used on some Ford and Chrysler (Mopar) models in the 1970–'71 era. The shaker hood consisted of the scoop mounted directly to the engine, peeking through a hole in the steel hood. The scoop was therefore above the stock hood surface, cold air would enter through the hood and the scoop would "shake" with the engine. Perhaps the largest production hood scoop was used on the 1968 426 Hemi Barracuda and Dart package cars.

Another unique scoop was the '70–'71 B-Body air grabber scoop, sometimes called the trap-door scoop, which could open and close from a switch in the driver's compartment.

Scoop Science—Basic hood scoop science says that the scoop should have about 30 square inches of area in the opening. Most of the hood scoops used on the 426 Hemi vehicles ('64–'71) fit within this parameter. The '64 and '65 scoops are wide and somewhat close to the hood. The '68 Hemi Super Stock A-Body scoop is huge.

The next basic advance was the Bauman boundary layer bleed-off scoops from about 1970, which move the scoop opening up off the hood surface and above the boundary layer that is next to the hood surface. These scoops were used on most of the early Pro Stock cars. The 1970 Challenger T/A scoop is also one of these and it was also used in Pro Stock (basic shape but taller). What is harder to define is the aerodynamic effect of the scoop. The scoop should be large at the starting line but small in the traps, so this is a compromise at best. The single four-barrel package is much more popular than the tall, tunnel-ram style manifolds, and this allows the overall scoop to be smaller or the package can even use underhood air.

A common test to run if you question the performance of your hood scoop or air cleaner: baseline your package with the air cleaner element installed. If you are checking the hood scoop function, then the air cleaner would be sealed to the hood scoop and the element removed.

Then you remove the air cleaner and rerun your test. If you are testing a hood scoop, then the seal to the hood scoop must be removed at the same time.

Note: A good hood scoop system should be good for 1.5 mph and 0.1 seconds in the quarter-mile. This means that the car should slow down by these amounts if you remove the hood scoop and go back to underhood air. Remember that if you tape over the hood scoop that you must remove the seal from the carbs to the hood so the engine can get underhood air.

Hood Scoop Air Cleaner—Each style of induction system and body style and hood scoop had its own style of air cleaner. The Street Hemi (two 4 bbls inline) used an oval one-piece unit that had two styles of base plates—a fresh-air version and an underhood version. The cross-ram Max Wedge or Race Hemi package had a unique one-piece but the individual ideal entrances with a screen across the top was the more popular setup. Many Max Wedge cars used two round air cleaners.

Drains—One of the unique features of the production fresh-air systems was that the base plates that connected the carburetor system to the hood scoop and sealed it to the hood had drains to allow any water that came in while driving it on the street to be directed away from the top of the engine and out of the engine compartment.

Hood Clearance—Obviously one of the functions of a hood scoop is to provide added clearance for the inlet system. While the details are unique to each intake system, engine and body style, many of the production scoops were added to power bulge hoods, which helped increase the space in the engine compartment. As the heads and intake manifolds get taller, there is no place for the carburetors to go except through the hood, which means a hood scoop, or power-bulge hood.

Cold Air—If you do not need any added hood clearance and you do not like the appearance of a large bulge on your hood, there are other ways of getting cold air to your engine. On a typical car, the air that the front end goes through is divided into three parts—the over-the-hood segment, the through-the-radiator segment and the under-the-bumper/radiator segment. In the last ten years or so, on MPI (fuel injection) cars, it has become popular to pick up some of the cold air that goes under the bumper and under the radiator. These systems use a long, molded plastic tube to connect the throttle body entrance to the air cleaner element that is relocated down next to the bottom of the radiator, on one side. One of these systems could be adapted to most carbureted engines but you might have to select the design carefully. This way you can have cold air without a hood scoop on the older performance vehicles.

Cooling—A V8 engine takes up a lot of space in an engine compartment. Large displacement engines and big blocks also make more heat. Headers have four exhaust tubes while an exhaust manifold has only one tube. There is more room around one tube than there is around 4 tubes. Getting hot air out of the engine compartment is not easy. If you can get the hot air out of the engine compartment, then the engine should run cooler and there is performance potential with cooler air.

Throttle Linkage Clearance—The vehicle's throttle linkage is a cable that runs from the firewall and accelerator pedal bracket to the rear of the intake manifold. If the carburetors end up pushed up into a hood scoop, you want to check the throttle linkage's clearance to the hood and the scoop. The specific throttle cable is unique to your vehicle, body and engine package but most are still available. Be sure that the cable has enough travel to open the carburetor system fully. Also remember to lube the cable on occasion.

CARB HARDWARE & INSTALLATION

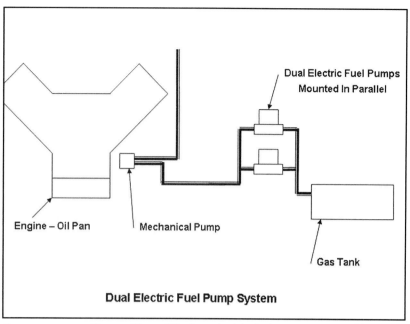

The carburetor uses a mechanical fuel pump in most production applications. The pump itself typically mounts to the side of the engine block and is driven off the cam. Today there are higher output aftermarket mechanical pumps from which you can select, similar to these Edelbrock units.

Drag-race cars will use either of the two electric fuel pumps mounted in parallel as shown in this drawing. Electric pumps can be used with the mechanical pump but if the mechanical pump is removed, then a fuel regulator must be added near the carburetor(s).

FUEL SYSTEM

The typical high-performance fuel delivery system uses an electric fuel pump mounted near the gas tank (at the rear), a mechanical pump mounted on the engine and a fuel pressure gauge plus at least one fuel filter. Many high-performance cars, like the 426 Hemi and 440 6 bbl, used a larger fuel line (3/8") than the standard engines (5/16"), so check yours. Typically the fuel line from the front to the rear is a hard (solid) line with braided-steel (flex) sections (available from XRP and others) on each end going from the body to the engine and from the fuel pump to the body. As the cars/engines get bigger and faster, there are either two electric pumps in the rear, or one very high-capacity pump (Barry Grant or Mallory) in the rear when the mechanical fuel pump is not used. In this second case, no mechanical pump, a fuel pressure regulator must be added at the front, near the carbs. The fuel system should also have a fuel filter mounted where you can see it and service it, but also protect it.

Another approach is to use two smaller electric fuel pumps mounted in parallel at the rear, no mechanical pump and a regulator at the front. With two electric pumps, you have a better chance of maintaining performance if you have an electric pump failure but today's pumps are very reliable.

Mechanical Fuel Pump

Most production cars used one standard mechanical fuel pump. In the last few years, the aftermarket has started offering high-volume mechanical pumps so you now have options.

An electric fuel pump is usually mounted in the rear of the vehicle near the gas tank.

Fuel filters come in many sizes and are available from many suppliers. You would like to have one close to the carburetor fuel inlet to protect it.

Electric Fuel Pump

With the gas tank mounted at the rear and the engine mounted in the front, there is a long fuel line connecting the two parts. The overall fuel delivery is often helped by using an electric fuel pump mounted in the rear, near the gas tank but protected and away from heat sources like the hot exhaust pipes.

Fuel Delivery

Most performance cars use an electric fuel pump in addition to the mechanical on the right front of the engine. If the mechanical is removed, then you must use a fuel regulator located near the carbs.

Rebuild & Powertune Carter/Edelbrock Carburetors

This is a vapor separator that also acts as a fuel filter. The single outlet (on left) goes to the carburetor. They were used in production on the 440 six-barrel and 426 Hemi engines. They are available from Year One. They are very helpful if you plan to operate the engine in hot weather.

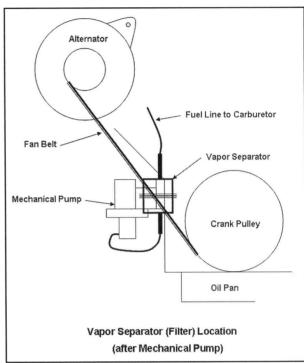

Vapor Separator (Filter) Location (after Mechanical Pump)

Generally you want to mount the vapor separator near the mechanical fuel pump which would mean low, on one side of the engine block, perhaps below the alternator and behind (rearward) the fan drive and pulleys.

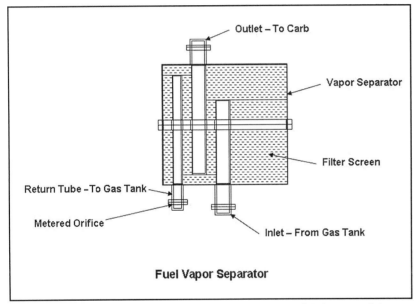

Fuel Vapor Separator

The vapor separator has three fittings: one inlet (bottom right), one outlet (top), and one metered orifice (lower left), which is the vapor return to the gas tank.

Fuel Line

I would recommend upgrading the standard 5/16" line used on most cars to a 3/8" fuel line. Check before replacing your existing fuel line because some cars, like the 426 Hemi and 440 6-bbl production cars, used the 3/8" fuel line as standard. On the other side of the coin, on cars that are over 35 years old, don't push your luck with an original fuel line—replacing it is cheap insurance.

Fuel Filters

If you have stock fuel filters in the chassis, you should replace them with new ones. Make sure the filter matches the size of the fuel line.

Vapor Separator

A vapor separator is a special kind of fuel filter. It is used to help keep high-performance engines from getting vapor lock. While most fuel filters have one inlet and one outlet, the vapor separator has three tubes—one inlet from the gas tank, one outlet to the carburetor and one return to the gas tank or charcoal canister. This filter should be mounted after the mechanical pump. Note that the inlet and the return are on the same side of filter and they should be on the bottom with the filter mounted in the vertical position. The top fitting (outlet) points to the carburetor. The vapor separator is very handy for hot weather racing and summer cruising. The 426 Hemi ('66–'71) and the 440 six-barrel engine packages used a special filter in production but it can be added to almost any engine package (available from Year One).

Fuel Type

The typical street engine should be constructed with about 9.0-to-1 compression ratio and therefore should use pump premium gas (92 octane). If you have a higher compression ratio, then you should use race gas. If you are planning on using an alcohol-based fuel like E85, ethanol or

Carb Hardware & Installation

There are many sizes and types of fuel pressure regulators. You generally want one that is designed for use with a carburetor and do not want one that was designed for EFI (electronic fuel injection) use.

Measure the fuel pressure just before the fuel enters the carburetor's fuel bowls. Be sure to mount the gauge so you can easily read it.

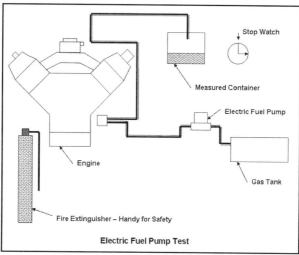

Electric Fuel Pump Test

If you use an electric fuel pump, then you should test the pump for performance. To do this, you need to time the pump for a given volume of fuel—how long it takes to pump a quart!

methanol, be sure that the hardware used in the fuel systems, seals, flexible line and filters is compatible with alcohol. Recalibration of carb jets will be required with alcohol fuels.

Fuel Regulator

If the mechanical fuel pump is removed from the engine and replaced by one or two electric fuel pumps, then you must use a fuel regulator. Fuel regulators are adjustable but they should be matched to the electric fuel pumps that are being used. For example, a electronic fuel injection system uses a fuel regulator that adjusts in the 25 to 35 psi range where a carburetor wants 5 to 7 psi of delivery pressure so the regulator should adjust in the 5 to 10 psi range. If in doubt, the electric fuel pump manufacturer can make a more specific recommendation.

Fuel Pressure

A typical carburetor performs best when fuel pressure set in the 5 to 7 psi range. This pressure should be measured just before the fuel enters the fuel bowls of the carburetor. The carburetor's needle and seat assembly is designed to operate at this pressure.

Fuel Pump Check

While not required on a street vehicle, I would recommend performing a fuel pump check on any fuel system that uses an electric fuel pump. Since most engine checks are run on a dynamometer, the fuel system used in these tests stays with the dyno. Once the engine is installed in the vehicle, the in-vehicle fuel delivery system has not been tested.

What you should do is test your actual as-installed fuel system on the vehicle in your shop. Since this is a performance engine, you need to record two things for the test—volume and time.

For volume, you will need a container that is approximately two quarts in size. Measure and pour in exactly one quart of water and mark the height around the container. Then discard the water and wipe it clean.

To measure the time, you will need a stopwatch or you can use the second hand on your watch. For safety, have a fully charged fire extinguisher nearby during this test, and make sure you know how to use it!

If you have a mechanical fuel pump, you can disconnect the main fuel line before it enters the mechanical pump and place this end in your container. Turn on the electric pump(s) and see how long it takes to pump one quart of fuel into your container. It should be in the 10–15 second range. If you have two pumps and it takes over 20 seconds for one quart, then you should test each pump individually. If you have a fuel regulator in your system, then disconnect the fuel line just before the regulator. If you have a mechanical fuel pump (standard) and want to test it, first disconnect the ignition by pulling the coil wire (so you can be sure that it will NOT start). Then disconnect the fuel line after the mechanical pump and before the carburetor(s) and turn the engine over on the starter. If the battery is in good shape, the engine rpm should be 550 rpm and at this engine speed, the standard pump should pump a quart in 1 minute (60 seconds). The caution with this test is that if the battery is tired (down on voltage), the engine rpm will be less than the specification and the time will go up but the pump might still be okay. A high-performance aftermarket mechanical pump would pump the quart faster or quicker.

Fuel System Plan

This can be a little tricky. The engine dyno guys will tell you that an engine will use about 0.5 lbs/hp-hr on gasoline. This is true at 350 hp and at 700 hp, and even twice these numbers. Getting an accurate horsepower number for your engine can be a little tricky, but you have to start somewhere. Everyone should know the cubic-inch displacement of their engines, and a common performance number for a good engine is one horsepower per cubic inch in output. A high-output, dual-purpose performance engine will make 1.5 hp per cubic inch. Using these two numbers, let's assume that you have a 350 cubic-inch engine. That means your engine could make 350 hp. If your engine is 500 cubic inches, then it would make 750 hp—high end (1.5 x 500). That means that our fuel delivery system should be able to pump between 175 lbs. and 375 lbs. of fuel in one hour.

Note: A fuel pump will pump more liquid in free-flow conditions—zero pressure—than it does at a rated pressure.

Regular gasoline weighs about 6.2 lbs. per gallon. Therefore, converting our weight calculation to volume of our specific liquid, our pumps should be able to pump 28.3 gallons per hour up to 60.5 gallons per hour. Since most street electric fuel pumps are rated at 70 to 75 gph at 7psi, you should have no trouble keeping up.

Refer to the Fuel Pump Check section on page 115 if you have an electric fuel pump that is rated at 70 gallons per hour. That means that it can pump 1.17 gallons per minute. Since there are four quarts to a gallon, that would mean that the electric pump could pump about one quart in just under 15 seconds (1/4 of a minute). Remember these are only general guidelines.

Chapter 6
Carb Adjustments & Related Hardware

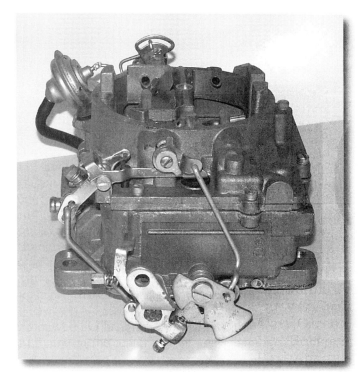

The left or driver's side of the AFB. This is a production version rather than an Edelbrock version. The main difference is the bowl vent, at the center of the photo, which is attached to the arm over the accelerator pump linkage. On this side, you adjust the idle settings, fast idle, accelerator pump linkage and part of the choke linkage.

Carburetor adjustments can be broken down into two main categories—those you should make before the carburetor is assembled and installed onto the intake manifold/engine and those that you can do once the engine is actually running. The preassembly/installation adjustments are based on estimates of whether the engine's A/F ratio is lean or rich. Keep in mind that there is more than one way to get to a specific A/F ratio so what you try to do is get close with the jet selection, which requires adjustments for your specific engine package, engine size and output estimates, and then fine-tune it with the metering rods because it can be done without further assembly.

PREASSEMBLY ADJUSTMENTS

Most carburetor adjustments are made after the engine has started and is running, with the carb fully assembled and running. If you're this far in your engine project, you do not want to take the carburetor apart. This means that you should have selected the jet size before you assembled the carburetor. For the AFB and AVS families, Edelbrock provides some basic guidelines for jetting selection or calibrations. In Chapter 4, I discussed how to estimate your jet sizes based on your specific engine displacement and estimated horsepower output. Without specific information, the idea is to keep the carburetor balanced. In other words, increase the secondaries as much as the primaries, or place half of the gain/loss in the secondaries.

Along with the jets, you should have set the desired float heights and installed the needle and seat assemblies that you

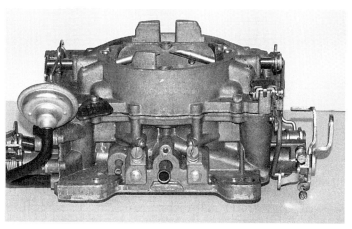

This is the front of a production AFB carb. The production AVS has only one idle mixture screw, so this example is more typical. The main adjustments on the front of the carburetor are the two idle mixture screws—left and right of center, toward the bottom. Edebrock AVS carbs also use two idle mixture screws.

want to use for your specific application. Changing primary jets, secondary jets, floats, float settings and needle and seat aspects, requires that the carburetor be disassembled. The better that you estimate these aspects, the less work and hassle you'll have.

At this point the carb should be assembled, and you are ready to install it on the intake manifold and engine. The next six sections should have been done based on previous

This is the right or passenger side of the production AFB. On this side, the choke is the main adjustment. The fuel comes in on this side but you just check that for leaks. The Edelbrock choke looks different than the production version but they function in very similar manner.

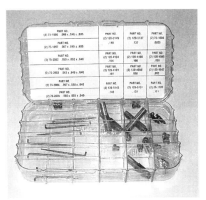

This is a Thermo-Quad jetting package called a Strip Kit by Carter. It included several Thermo-Quad metering rods, primary and secondary jets along with a key to decode what they were. These are no longer available new.

Without extra service parts, you might have to resort to drilling out your existing jets. Use numbered drills to determine the exact size of the orifice if you can't read or decipher the codes on the jet itself. This is a Thermo-Quad primary jet. Do not drill out any jet until you have a spare like the original, just in case.

chapters. Now is the time to complete or double-check these items. Following these six topics, the next eleven topics listed under the "General Checks" heading are reminders that these items should have also been done in previous steps. Remember that once the carb has been installed, everything has to be taken off, drained and disassembled at a later time, which is a lot of extra work. Performing double-checks now will save time and eliminate future problems.

Jetting Packages

Edelbrock offers a calibration kit for tuning the AFB and AVS carbs that features several selections of metering rods and main/secondary jets. Carter also offered tuning kits for the Thermo-Quad called Strip Kits, which had many main jet, secondary jet and metering rod selections. You want to have these basic tuning pieces in hand before you start the actual calibration tests in Chapter 7.

Fuel Selection

Don't forget that fuel is very important to how the carburetor runs. The obvious ones are alcohol and E85 because they change the basic fuel setting in the carburetor and old fuel, whether it has been sitting in the fuel tank for a year or a less-than-half-full 55-gallon barrel, because it has no octane left. It might start but the low octane will affect the actual fuel setting, so you don't want to fine-tune with it. What isn't so obvious is the gasoline itself. There are many different levels of gasoline. The obvious one is pump gas versus racing gas, but there are several manufacturers of racing gas and each company offers six or more levels of racing gas. They are rated by their octane capability but they will also affect the fuel curve that your carburetor delivers to your engine, so this has an effect on your basic tune-up. Do not perform your tune-up on one level of gas and then switch to another.

Gasoline—There are a lot of different blends of fuel that offer different features. Usually the gas is selected based on the engine's compression ratio. The type of racing can also enter the equation. The compression ratio dictates the fuel's octane rating. Pump gas has an octane rating of 92 and therefore the use of pump gas limits the engine's compression ratio to 9-to-1. Racing gas offers much higher octane ratings, as high as 115 but 105–110 is more common, and is much more expensive. A 10- or 11-to-1 compression ratio engine can probably use the high octane racing gas mixed with pump gas—perhaps 50/50. There are several manufacturers of racing gas and each one has their own coding system. Most race tracks will perform a fuel check at the track and most of these fuel checks are based on the fuel's specific gravity. So the specific gravity is also listed on the manufacturer's spec sheet. Most common racing fuels (gasolines) range from 0.700 to about 0.740 on the specific gravity scale. The lower number tends to be the higher-octane versions. You need to know the fuel octane rating and its specific gravity, and you would like to know the viscosity, but that number is not published. All of these things contribute to the engine's fuel curve and calibration. So once you have done your calibration with a specific brand and blend, do not change fuels.

Note: Racing fuel (gasoline) is available from VP Fuels, Shell and Sunoco.

Carb Adjustments & Related Hardware

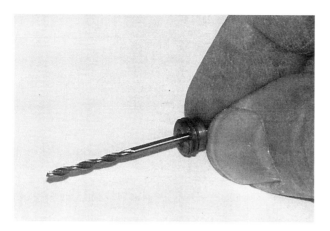

This is an AFB/AVS jet. You can use a numbered drill to double-check the jet size based on the number on the top surface. Edelbrock services many sizes of these jets so drilling them out isn't required. The tall, three-step metering rod jet used in the primary side of some production AFB and AVS carburetors is not as readily available. One solution would be to switch back to the two-step system—metering rods, jets, springs and flat covers. Another would be to drill out the jets. Do not do any drilling unless you have spare parts.

This is a production velocity valve. Not all production valves looked like this. Many looked like the unit shown at right from the smaller cfm AFB.

This is the velocity valve typical of what is used in the AFB carburetors. The weight that appears added to the upper counterweight is the amount that would be cut off of the counterweight if you wanted to make the secondary velocity valve open quicker, which might be desired in a quicker drag-race car.

Tip: The most popular rating for gasoline is octane. Technically the rated number given should be $(R + M)/2$. The R and M stand for two different methods of rating gasoline octane. R stands for Research and M stands for Motor. Generally the R or Research number is higher than the Motor number so the two numbers are added together and then divided by two to get an average.

Generally the fuel manufacturers add color to the various blends of fuel so you can tell one from the other. Blue, yellow red, green or purple are just dyes that are added to the basic fuel for identification and are not the reason that the octane rating or specific gravity changed.

If you have a specific gravity gauge, it can be helpful for determining if your fuel is still good. If you bought a 55-gallon drum of racing gas that had a specific gravity of 0.715 at the time of purchase, and it is now 0.760 or higher, you might want to think about not using it! With any fuel, the light ends evaporate first and this makes the fuel get heavier. The light ends are also directly related to the fuel's high octane rating, so as they evaporate the fuel's octane rating is dropping too. Use caution and check with your supplier if you find this situation.

Jets

The primary jets and secondary jets were installed prior to the final assembly of the carburetor. There are identification numbers etched into the top surface of the jets and you should have these written down. From here on, you should plan on using the metering rods to change the calibration.

Velocity Valve

Any adjustments to the velocity valve on the AFB should have been done before the carburetor was assembled. The AVS version has the spring-loaded air door, which you can adjust after assembly—see later section.

Floats

Like the jets, the floats should have been set at assembly. To change a float setting now requires the carb to be disassembled.

Air Cleaner

This isn't as important as some aspects that relate to the fuel curve, but the air cleaner will affect the A/F ratio delivered to the engine. The installation of an air cleaner will richen up the engine or if the air cleaner is removed, it will lean out the engine.

General Checks

In the general step-by-step reassembly process for each carburetor detailed in Chapter 4 there are several steps that would fall into the checking category more than actual adjustments.

Additionally these checks and adjustment items are

This is the bottom view of the AFB-AVS carburetor. The primaries are toward the top. Note that the primary throttle blades are not quite closed fully. Pay very close attention to the two small holes just above the tip of the primary throttle blades.

This is the bottom view of the Thermo-Quad carburetor. The primaries are toward the top. Carefully observe that the primaries are not fully closed. There are two small holes just above the tip of each primary throttle blade. The small slot that is just above the tip of the throttle blade is the transfer slot. If the throttle blade was fully closed, this slot should be covered. The small hole above the transfer slot is the opening for the tip of the idle mixture screw. This is the orifice that feeds fuel to the engine at idle and with the throttle fully closed. Remember that the carb is upside-down and the idle hole is now above the throttle blade but actually would be below in its normal position.

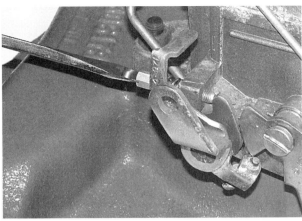

The basic idle screw is easy to reach and adjust. You would like to start with the throttles fully closed. The fast idle cam and adjusting screw tends to be on the bottom of the linkage—toward the lower right. This makes it very difficult to adjust once installed but can be done with the engine shut off by opening the throttle. It is a lot easier to understand if you check out this part of the linkage while the carb is off the engine.

When the carburetor is off the engine/manifold, the throttle linkage should allow the throttles to close fully. This is a Thermo-Quad and the primaries, toward upper right, are slightly open. You want to start with the throttles fully closed before you install the carb onto the engine.

listed in three different sequences based on the specific carburetor and/or source. So I have made a list of these items that would seem to work for our somewhat unique universal applications.

Secondary Throttle Linkage—The key to this step is to make sure that the secondaries actually open and can open smoothly. This basic clearance check is detailed in Chapter 4.

Choke Operation—There are several items that are considered part of the overall choke operation. There are specific aspects that relate to the manual or thermostat controlled choke or an electric choke. Depending upon the style there are the choke control lever, the vacuum kick, the choke unloader, choke diaphragm connector rod, and the choke index, but as a sum total, you want the choke blade on top of the primaries to close when cold. If it does, then the engine should start (troubleshooting is in the next chapter). Once started, the choke should open or pull off but any adjustment of this aspect requires the engine to be running so that is in the Post-Assembly section on page 122.

Fast Idle Cam and Linkage—Again, the fast idle cam and linkage are shown/discussed in Chapter 4. The purpose of this linkage is to have the engine idle at a higher engine rpm when it is cold and to drop back to its normal engine idle rpm as the engine warms up. The key to this step is to find the

Carb Adjustments & Related Hardware

This is an AFB showing the accelerator pump linkage—left side of the carburetor. The pump rod goes from the bracket on the bottom of the linkage upward to the pump arm that extends down from the center pivot. There are 3 holes in the pump arm. This can be adjusted if more or less accelerator pump is desired. It is shown in the middle hole. Moving the rod closer to the pivot creates a longer pump shot.

There are generally two idle mixture screws in each carburetor with the exception being the production AVS which only has one and it has a left-hand thread on the screw. While similar, the AFB and Thermo-Quad mixture screws are not the same. Both do have the small spring that goes on each screw—lower left and lower right. The AFB and AVS mixture screws are generally visible and accessible for easy adjustment but the Thermo-Quad mixture screws have a black plastic cap over the screw heads and it must be pulled off before you can adjust properly.

This is the Thermo-Quad showing the accelerator pump linkage. It also starts at the bottom and goes upward to the pump arm but there are only 2 adjusting holes. It is shown in the closer hole to the pivot.

actual adjustment screw and fast idle cam because the screw tends to be on the bottom of the linkage and not as easy to see once installed on a manifold. To fine-tune this step you have to wait for the Post-Assembly section beginning on page 122.

Idle Speed—The key to this step is to be sure that the primaries are fully closed. In some cases, the idle adjusting screw holds the throttle blades partially open and this will make adjusting the idle more difficult.

WOT Opening—This step actually has two parts. The first is to check the WOT (wide open throttle) stop on the carburetor itself as detailed in Chapter 4. The second part is to check this position once the carburetor is installed on the engine and in the car. It does require two people—one in the car to push on the accelerator pedal and one to check the WOT position on the carb by checking the throttle stop and looking down the throttle bores—use flashlight if required.

Bowl Vent Valve—Only the production AFB and AVS carburetors use a bowl vent. The arm sits over the accelerator pump arm and opens the vent when the linkage is closed. This can be done at assembly or after installation.

Vacuum Throttle Positioner—This device is only used on the production Thermo-Quad carburetors built for or sold in California. To adjust properly, the engine has to be running (next section), but you should know where it is and check to see if it is lined up correctly—Chapter 4.

Accelerator Pump Linkage—The key to this step is to be sure that the accelerator pump functions properly and that the linkage is set in the middle hole. Prior to starting you will also check the accelerator pump for basic function by turning the electric pumps on and filling the float bowls or by turning the engine over on the starter and allowing the mechanical fuel pump to fill the bowls. Once

When adjusting the idle mixture screws, an AFB shown, it can be helpful if you place a piece of masking tape or black mark on one side of the blade to help enable you to count full revolutions.

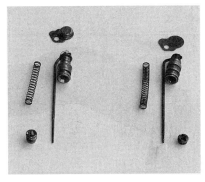

The two-step metering rod system is on the right and the three-step system is on the left. They can be swapped as complete systems. The bottom step on the metering rod, closest to the bottom of the photo, is the smallest step and is the step that determines the WOT fuel in conjunction with the primary jet.

there is fuel in the bowls, you can use the accelerator pump to squirt fuel into the intake system. With the air cleaner off, you can watch this procedure and confirm that the basic pump operation is acceptable.

Check for Leaks—At this point, it is a good idea to stop and check for fuel leaks on the carburetor itself and the fuel pumps and fuel line and all fittings.

Air Door Spring Tension—The air door tension is only used on the AVS and Thermo-Quad carburetors. The two mechanisms are slightly different (see Chapter 4) but both should be initially set to 1 1/2 to 2 turns of spring tension. This adjustment is done at assembly and will be fine-tuned after the engine is running. On the Thermo-Quad, the second aspect of the air door is the amount of opening. Initially this amount of opening only needs to be measured. I'll discuss further adjustments later in this chapter.

Idle Mixture Screws—There are two different basic systems—the Edelbrock, Thermo-Quad and AFB versions which use two right-hand thread mixture screws and the production AVS which uses a single left-hand thread adjuster. In either case, adjust the mixture screw(s) to 1 1/2 turns off the seat (closed position) as the initial setting. Further adjustments will be done in the next section.

POST-ASSEMBLY INSTALLATION & ADJUSTMENTS

For this section we'll assume that the engine is running. Basic troubleshooting dealing with non-starting problems and such will be discussed in detail in the next chapter. While listening to the engine, note how smoothly it runs; looking at the tach closely for specific rpm readings is very important. For fine-tuning the idle steps you should use a vacuum gauge. As you adjust the idle fuel mixture, try to get the vacuum gauge to read the highest number.

It is very, very important to remember during carburetor tuning sessions to break down the engine performance into three areas or three speeds—idle, part throttle and wide-open-throttle or WOT. Address these areas one at a time. That's one reason the initial estimates are so important, especially on jet sizes. For example, if you missed the jet size and have to change, the jet change affects the idle, the part throttle and the WOT. Therefore everything that you did initially has to be redone. One advantage of Carter/Edelbrock carbs is that there is more than one way to change the A/F ratio in the engine, so a jet change may not be required even if the guess is wrong.

In this section, I will discuss what the various adjustments are, the mixture screws, the jets and the metering rods, and what they do and how to adjust. Basically carburetor adjustments are used to try to solve A/F ratio problems with the engine. So you must estimate your engine's performance at each speed compared on the A/F ratio—is it lean? Is it rich? The lean/rich answer gives you the direction for your adjustment.

Idle Mixture Screws

The Edelbrock carburetors, the Carter AFB (production) and the Thermo-Quads all use two idle mixture screws. The Carter AVS (production) used a single mixture screw and it used a left-hand thread. On the normal right-hand thread mixture adjusting screws, turning the screw clockwise leans the idle mixture, while turning the screw outward, or counterclockwise, makes the idle mixture richer. This process is reversed for the left-hand thread production AVS single screw—clockwise (outward) makes it richer and counterclockwise makes it leaner. In both cases, the initial recommended setting is 1 1/2 turns from fully seated (Chapter 4). If the initial engine idle information is that the engine is too lean, then go to 2 turns or 2 1/2 turns on each side, while watching the vacuum gauge.

Note: The idle passage is fully open at about 3 turns off the seated position. At that point further adjustment will not change the idle A/F ratio.

Metering Rods (Step 1, WOT)

The smallest step on the metering rod is the WOT (wide-open throttle) step. The latest service

Carb Adjustments & Related Hardware

Rather than two separate metering rods, the Thermo-Quad as shown has an arm that extends across the carburetor and connects to two metering rods. It is always a good idea to check this linkage out by pushing down on the arm with a finger over each metering rod well.

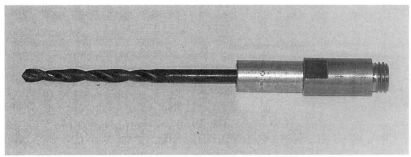

The Thermo-Quad secondary jet is unique because it is tall and thin. The flat on the side about halfway from each end is designed for wrench removal. Decoding/reading the stamped/etched numbers on the jet may not be easy and you should check the orifice size by using number drills. Service parts may be hard to find and you might have to drill out the jet to increase the fuel flow, but don't drill unless you have spare parts.

Another unique feature of the Thermo-Quad carburetors is that the basic height of the metering rod arm can be manually adjusted by a small screw located in the top of the piston and accessed through a small hole in the retaining bracket.

hardware from Edelbrock shows many, many metering rods available but the WOT step runs from 0.037" to about 0.057". With a jet, the bigger the orifice, the richer the package is but with metering rod it is reversed—the smaller the metering rod diameter, the richer the mixture and the larger the step, the leaner the mixture. During the assembly stage in Chapter 4, I suggested using a 0.047" metering rod because it is in the middle of the adjustment range. If the engine's WOT performance indicates that it is too rich/lean and the metering rods have been adjusted all the way in that direction, then a primary jet change is indicated and this should be done early in the adjustment sequence, to save time and work.

TQ Metering Rods—The Thermo-Quad carbs have an added adjustment to the metering rods for the standard AFB and AVS models. The TQ uses one center piston with a rod that goes out to each metering rod—one on the left and one on the right. The central piston which moves the two metering rods up and down as the engine goes from part-throttle operation to wide-open throttle operation has an adjustment on top. The bracket that holds the piston and arm assembly in place has a small hole in the top which allows a small screwdriver access to the screw. This screw controls the position of the metering rods. Rotating the screw clockwise richens up the mixture. Turning the screw counterclockwise lowers the rod position and leans the mixture out. This adjustment will affect part-throttle performance not wide-open throttle performance, so it can have an effect on both fuel economy and drivability.

Primary Main Metering Jets

The idea is not to change the jets based on a good estimate. Currently, Edelbrock offers jets from 0.077" to 0.119". Large jets make the engine richer, but each step is a pretty large one—about 8%. Each jet size change is about 0.003", for example 0.095" to 0.098". If you change a primary jet, then go back to the middle metering rod so you can make fine adjustments again. The Thermo-Quad jets are unique and not currently available but they were common in the 1970s and '80s, so you might find some used. The AFB uses the same jet in the primary and secondary locations. Some of the production AFBs and AVSs used a 3-step metering rod package. This meant that they had taller primary jets and the standard or short secondary jet. While readily available in the past, these tall jets

An AFB/AVS accelerator pump nozzle is in the center, the secondaries are toward the bottom. Each nozzle points toward the center of adjacent primary.

A Thermo-Quad accelerator pump nozzle in the center, secondaries toward the bottom. Each nozzle points toward the center of adjacent primary. Since the Thermo-Quad's primaries are farther apart, these nozzles have a greater angle between them than the AFB/AVS version—pointing almost straight out.

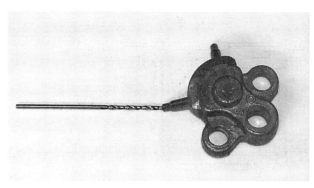
This is an AFB/AVS accelerator pump nozzle. Since Edelbrock has at least 3 different sizes of nozzles available, I would not recommend drilling out the nozzles. The Thermo-Quad nozzles are not serviced, but generally a bigger nozzle is not going to help much on these carbs.

are no longer serviced. If you have a 3-step metering rod package and want to convert to the more readily available 2-step system, you must change the jet, the metering rod and the cover.

Secondary Metering Jets

Since the secondary jets can use the same pieces as the primary ones, they have the same range of jet sizes available: 0.077" to 0.119". Keep in mind that the secondary jet size has very little effect on the idle and part throttle aspects of the engine operation. Therefore, if the engine performance indicates a jet change, either rich or lean, and the idle and part throttle aspects seem fine, then it might be best to make only a secondary jet change, and leave the primary alone. The Thermo-Quad uses unique jets—real tall. The Edelbrock jets will not work in this application. Since the Thermo-Quad carburetors were built during the emissions era, it would be very unusual to require less fuel than the stock setup. If you require more fuel, the only solution may be to drill out the existing jet to the next size. I would strongly recommend that if this is the case, that you get several spare carbs off the Internet to use for spare parts.

Metering Rods (Step 2, Part Throttle)

There are 2 steps on the standard metering rods. The bottom/small step is for wide-open throttle. The second step is larger than the first. The current service parts range from an intermediate step of 0.057" to 0.075" in increments of 0.001" to 0.002", so there are a lot to choose from. This test has to be made under load and that means that you have to drive the car. If you have an automatic transmission, then the loads are higher in third gear than they are in first. In your driving test, try vehicle speeds in the 25 to 60 mph range and try to be consistent. Watch the rpm and the vacuum gauge if you have one installed. Try to estimate whether the engine is rich or lean. The lean condition is easier to spot because it will yield a lean/surge response from the engine, which is quite easy to feel when you're driving.

Note: The hardest part of any calibration is the part throttle phase. Once you have the two ends of the curve tied down—idle and full throttle—then you have a much better choice of fine-tuning the part throttle part.

Accelerator Pump Linkage

At the rebuild stage in Chapter 4, I suggested that the accelerator pump linkage be installed in the middle hole. In some cases there are only two holes, so you pick the hole farthest from the pivot. Either of these setups is fine for street use. In either case you have an adjustment in the form of a hole closer to the pivot. Typically you don't want to move the linkage to the closest hole until you go drag racing. The standing start, max acceleration contest on drag racing slicks (rear tires) tends to require more fuel at launch and that means more pump shot, which comes from moving the linkage closer to the pivot.

Accelerator Pump Nozzles

Generally this is not required but Edelbrock does offer larger pump shooters or nozzles for the AFB/AVS versions. Remember that a larger diameter nozzle squirts more fuel quicker so the shot does not last as long. This may be better for very large displacement engines. A smaller nozzle

Carb Adjustments & Related Hardware

The AVS uses an air door over the secondaries and it is located at the top of the carburetor. The air door opening is controlled by a spring. The tension on the spring is controlled by turning shaft that the air door pivots on and then locking that position in place. This is the Edelbrock version of the AVS. The large screw in the center is the end of the shaft that you turn to increase or decrease the amount of tension on the spring. The smaller screw, slightly above and to the left of the big screw is the locking device. Edelbrock uses a Torx screw in this location.

On the production AVS, you can use two screwdrivers at the same time. Loosen the locking screw with one while holding the notch/center screw in place. Then adjust the center screw/notch to the desired tension. If you just loosen the locking screw, then the tension on the spring releases quickly, and you have to start from scratch. The Edelbrock version can be done with a Torx and a screwdriver.

The Thermo-Quad carburetor uses a similar system on its air door but it is somewhat harder to adjust. The locking screw is just to the left of the choke linkage in the center of photo. The adjusting screw is recessed inside the outer locking screw. There is a special tool but they are expensive. A large-blade screwdriver can loosen to outer locking screw. Then the inner, smaller screw can be adjusted for the desired tension. Holding this tension, a small screwdriver can be used to tap tangentially at one of the two slots in the outer locking screw and tighten it enough to hold the tension while you switch to the big screwdriver and lock it in.

With the big, wide, thick-blade screwdriver in place to loosen the locking screw, there is no room for the smaller screwdriver to hold the existing tension.

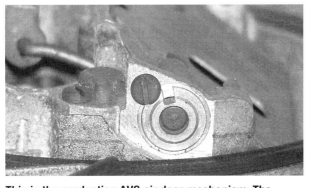

This is the production AVS air door mechanism. The locking device is the large screw in the center of photo. The tension adjustment is slightly lower and to the right but has a notch and smaller screw recessed inside adjuster. Loosen the locking screw and then turn the center notch or screw. Hold the notch as you tighten or loosen.

would give you fuel for a longer time and this might be advantageous for very small engines or cars that are racing on real street tires or engines with really heavy flywheels (manual transmission versions).

Main Idle Set or Throttle Stop

Many people like to adjust the idle by opening the main throttle plates by adjusting the idle screw. A slight amount is fine once the idle mixtures screws are adjusted. For example, if you desired an idle of 850 rpm and it was currently at 700 rpm. However, what typically happens if you open the primary throttles as part of your idle setting is the in-neutral idle will be too high, perhaps 1200 to 1500 rpm and the in-gear idle (assuming an auto transmission) drops to 750 to 850 rpm. This means that the idle setting is too lean. To richen, turn out the idle screws.

Secondary Air Door

AFB carburetors do not have an air door, so this section does not apply to them. Both the AVS and Thermo-Quad carburetors have a spring-adjusted air door over the secondaries. At rebuild, I recommended that you use 1 1/2 turns from the no-tension position as your basic starting point. More tension means that the airflow effect of the secondaries is delayed, while less tension means that the effect occurs more quickly. The best setting may change from the street to the strip. Large

REBUILD & POWERTUNE CARTER/EDELBROCK CARBURETORS

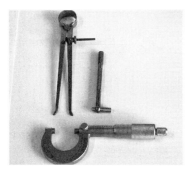

Before you adjust the Thermo-Quad air door, you must learn how to measure what the air door opening is. The first step is to gather your measuring devices. You'll need a 0 to 1" micrometer (bottom) and a snap-gauge (upper right), with 1/2" to 1" measuring capability. The inside calipers—upper left—are probably the easiest to use.

The standard service procedure is to use a steel scale and hold the air door fully open with one finger, and use the steel scale to measure from the tip of the top edge of the air door to the choke well wall—straight. Quick and easy.

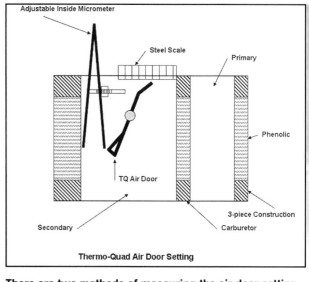

Thermo-Quad Air Door Setting

There are two methods of measuring the air door setting and therefore two specifications. The steel scale is used to do the service procedure from the top edge of the fully open air door to the choke well wall. The racing or performance procedure (the one that I use) measures from the rearward tip of the fully open air door to the black plastic wall in the secondary. The black plastic wall is straight, flat and vertical in this area so this can be a very accurate measurement. Because it is down inside the carburetor, it isn't as easy or as quick as the service procedure.

I prefer to measure the air door setting from the rearward tip of the fully open air door to the black plastic wall of the secondary—as shown here. Hold the air door fully open with one finger and measure using the snap gauge. It is probably easier to set the desired number with the snap gauge in the micrometer and then transfer this to the carburetor to see if the current air door setting is greater or less than your desired setting.

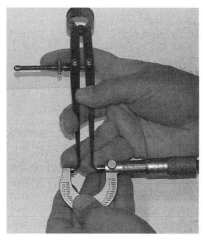

Using the caliper, you can set the correct or desired amount first using the micrometer. Or you could actually measure the air door opening but it is a little tricky to hold the air door open with one hand and adjust the caliper with the other while holding the caliper in the proper place for the measurement.

To adjust the amount of the air door opening on the Thermo-Quad carburetor, you first must find the small foot that sticks out from the left-side metering rod cover plate—horizontal arrow. And then find the slot in the top of the air door—vertical arrow.

displacement engines may want them open sooner while small displacement engines may want them to open later. The chassis (gear ratios, converters, weight, suspension, etc.) may affect this as well so it is best left until last, and realize that if you develop the best opening for your specific package on the street, you may want to change it if you go drag racing.

Air Door Opening—The AVS air door opens straight up so the amount of opening does not apply to any of the AVS carburetors. However, the Thermo-Quad has an adjustable amount of opening on the air door itself. It can be adjusted with the carburetor fully assembled. It can be adjusted with simple tools—a pair of pliers. Perhaps the difficult part is to discuss how to measure the setting. There are two basic techniques. The first is to use a steel scale and measure from the top of the air door at fully open to the choke well wall at the center of the carburetor. This method is recommended by the production service procedures. The other method is to measure from

Carb Adjustments & Related Hardware

Almost fully open the air door and using a pair of pliers, bend the tab in the desired direction.

Rebend the tab if the new setting isn't where you want to be—yet. Repeat until air door setting matches the desired number.

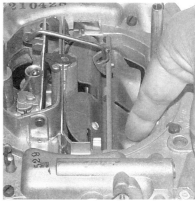

The proper procedure for setting air doors is 1, 1 1/2, 2 or 2 1/2 turns on the tension spring. It is always a good idea to test the spring tension after you are done setting it by poking at it with your finger. You poke downward at an angle and aim at the tip and watch to see how it bounces. With practice, this quick test checks the basic function of the air door and estimates the tension at the same time.

After the tab is bent, then use your calipers to remeasure the new setting.

Once the carburetor is fully assembled, check to be sure that the choke can close fully—as shown. The primaries are toward the top.

For normal operation, you want the choke blade to come off fully, which means that it should sit completely vertical—as shown. Primaries are to the right.

the tip of the rear segment of the air door at fully open position rearward to the black plastic wall in the secondary. The typical service measurement is about 0.500". The typical performance or racing number (second method) is about 0.850". If you want to change the amount of air door opening, there is a tab or slot on the leading edge of the air door on the left side (throttle side). The slot is about an inch long and runs parallel to the centerline of the shaft. The outer end of this tab, which is created by the slot, acts as the stop for the air door in conjunction with a small foot that sticks out from the left side metering rod cover plate. By bending the tab you can change the amount of air door opening.

Velocity Valve

The velocity valve in the AFB carbs can't be adjusted without taking the carb apart. However, you can evaluate its performance—check to see if it opens too soon or too late—the same way as you do

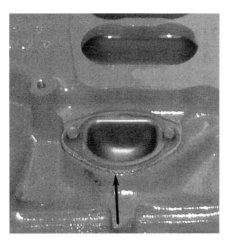

On production carburetors that used a mechanical choke, which are actually thermo-mechanical devices, the bi-metal coil spring changed its length with temperature and sat in the choke well on the production intake manifolds. This choke well sat on top of the heat crossover passage that was cast into the bottom part of the intake manifold. This choke well in the manifold was generally covered with a stainless steel cover (arrow) or cup, which was held in by two screws. The same two screws would also hold the bi-metal coil spring.

the spring-loaded version used on the AVS described earlier.

Choke

There are several versions of chokes. There are manual chokes and electric chokes (Edelbrock) and in many cases you can select one or the other. In either case, once the engine has started and is running, the choke can't be adjusted until the engine is cold again, so this is usually done in the second round of tune-ups. While the details change, you basically want the choke blade to be closed when the engine is cold and you are just trying to start the engine. As the engine warms up, you want the choke blade to go vertical.

OTHER ADJUSTMENTS

Carburetors are adjusted for many reasons. While almost everything in the engine will affect the carburetor circuits, there is a misconception that it is only affected by the actual running of the engine and its associated hardware. I'll discuss the hardware issues, both direct and indirect, in the following sections, but there are a couple of very important factors that do not fit in either of these areas.

Weather

Weather conditions are very important to the engine and the carburetor. There are four main factors: ambient temperature, air density, humidity and altitude/barometer. There is a giant mathematical equation that is used to calculate the detailed effects of weather. This equation is in many dynamometer software packages used for engine development. It appears as a correction factor. If you test an engine on a cold day, it will make more power than it does on a hot day, so the correction factors allow you to compare one days runs against another run under different conditions. Engines also make more power at sea level (zero elevation) than they do at elevation, so correction factors allow you to compare performance numbers from two different locations.

Altitude is related to the barometer, so when the weather forecaster predicts a high pressure center is coming, the engine's power will be up. Low-pressure centers hurt performance.

Racers probably change the A/F ratio in their engine more often because of weather conditions than any other cause. Street engines are not usually changed once they are set up properly. To do this, you have to have a weather station that measures the air temperature, the barometric pressure, and the relative humidity.

Remember that one jet size in the Carter/Edelbrock carburetor translates into an 8% change in A/F ratio. However, while the steps from one jet size to another stay at 0.003", the actual percentage gain for 0.077" going to 0.080" is not the same as from 0.116" to 0.119", but I don't want to turn this into a mathematical course so let's stick with 8%. Changing the metering rod offers changes in the 2% to 3% area. So a 2,500-foot change in elevation might require a metering rod change while a 5,000-foot change (LA to Denver) might require a jet change.

Engine Condition

The condition of the engine might sound like a hardware question but it really relates to the shape that the hardware is in rather than the hardware itself. The first question is, does it burn oil? If oil gets past the rings, goes down the valve guides and slips past the gaskets, usually head gasket or intake gasket, then this oil is burned in the combustion process. You will notice smoke coming from the tailpipe or a higher than normal usage of oil. Engine oil is not a good fuel compared to gasoline so if the engine is burning any amount of oil, this affects the engine operation and can cause problems that relate to the engine's idle, part throttle or max power performance.

There are specific tests that can be performed on the engine like the compression test and the leakdown test. Both of these are detailed in Chapter 7. These tests can help you determine the condition of the engine, new or old, even if it does not burn oil. They can also help you determine which

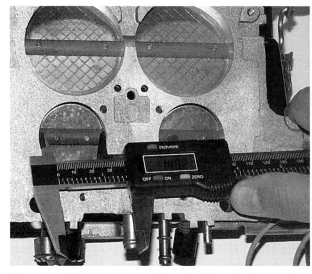

It is a good idea to measure the actual throttle bores of the carburetor that you are going to use. Most carburetors today are rated by cfm so you want to connect that to the actual throttle bore size.

Carb Adjustments & Related Hardware

There are many styles of air cleaners such as this basic 14" round unit.

Production-style air cleaners like this dual-snorkel unit are also possible.

For racing, you could use an ideal entrance like this one with a screen over the top.

The econo ideal entrance can be made from an old production air cleaner bottom. Remember that you need to secure the base in position or it will fall off with hard acceleration.

A hood scoop like this shaker unit is very important to the carburetor adjustments because the calibration should be richer with cold air or no element.

cylinder or cylinders are the main cause of your problem(s) if you have one.

DIRECT HARDWARE FACTORS

Almost every part that goes into an engine can have an effect on the carburetor. There are very few parts to the engine that do not affect the overall A/F ratio at one of the engine speeds—idle, part throttle or WOT. Deciding which hardware fits into the direct category and which fits into the indirect category was not so easy. My dividing line was that the carburetor bolts directly to the intake manifold and the air cleaner bolts directly to the carburetor as does the fuel line, so these items fit into the direct category. The rest of the engine's hardware fits into the indirect category.

Dynos

Dynamometers can be very expensive but also very useful. Dynos can be used for calibration and fuel curve work but it is not mandatory. The in-car calibration just takes time and patience and a safe place to do your testing.

Engine Variables

There are many items that fit into the engine variables category like spark timing, valve lash, valve timing, fuel pump pressure, compression ratio, exhaust back pressure, airflow and cam events. Many of these variables are the result of the effects of more than one part. So rather than discuss the engine variable, I will discuss the hardware.

Inlet System

The inlet system generally includes the carburetor and intake manifold but for this section I am referring to how the air gets to the inlet of the carburetor. The easy answer is through the air cleaner, but it is important to include the hood scoop or cold-air inlet too. You should do your tuning steps with your air cleaner installed. This is important because the installation of an air cleaner richens up the A/F ratio that the carburetor delivers to the engine or the removal of the air cleaner leans it out. Additionally, if you have a functional hood scoop that provides cold air to the carb inlet, or one that provides forced air, then the jetting package will have to be richened up for the more dense (cold) air.

If the engine is being raced, it is common to remove the air cleaner and use an ideal entrance or velocity stack. These can be fancy ones or just the base of the air cleaner but the ideal entrance helps smooth the airflow going into the top of the carburetor. The typical round air cleaner, 12"–14" in diameter, tends to be open on the face of the element. Production air cleaners enclose the element and add a snorkel. This allows the production engines to have a heat stove which helps provide warm air from around the exhaust manifold quickly which helps with cold drivability. The next step is the dual-snorkel air cleaner and then connecting one or both of the snorkel tubes to a cold-air entrance usually in the grille area. As the

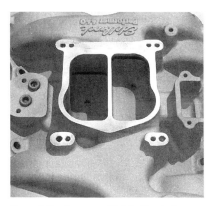
One of the key manifold effects is related to the basic design—this is a standard aluminum dual-plane and they tend to be good street manifolds and make good torque.

This is a standard single-plane. Single-plane manifolds are generally used in racing. They tend to make more power but less torque and this can cause problems for the carburetors if used on the street or in dual-purpose applications.

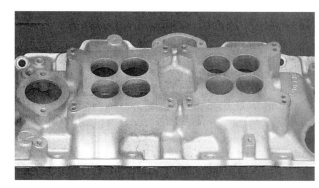
This is a single-plane, eight-barrel inline manifold and is probably best used in a restoration application. Low-speed torque is going to be limited and this can be a challenge for the calibration.

This is a race dual plane which can be used in both street and strip applications. They tend to be much taller or higher than standard manifolds so check the hood clearance.

This is a new dual-plane eight-barrel inline and is much better suited for street or dual-purpose applications.

This is a race single-plane manifold with larger runners that extend into the plenum area. These manifolds are designed for racing and this means more engine output and higher rpm, so the carburetor has to be adjusted for these conditions.

inlet system provides more air and colder air to the engine, the engine will make more power and the carburetor has to deliver more fuel.

Manifold Heat Crossover Passage

Most production-based cylinder heads on modern V8 engines have a heat crossover. There will be a mating passage in the intake manifold. These heat crossovers help with the cold drivability. For summer use and drag racing applications, it is common to block these heat crossovers. This is considered making the manifold cold. For racing purposes, the cold manifold tends to make more power but blocking the heat in the manifold can affect the manifold's fuel distribution, which in turn can decrease engine output. Typically, cold intake manifolds require more fuel from the carburetor.

Fuel Distribution

Another aspect of intake manifolds is the actual fuel distribution. This aspect deals with the A/F ratio of each cylinder. Ideally, they would all be the same but this is rarely true. You can have one or two cylinders that run hot (lean condition) and/or you can have some that run cold (rich condition). Generally the fuel distribution of an intake manifold-engine-carb combination is done on a dyno. Typically it requires the headers or exhaust manifolds to be equipped with special temperature sensors that will tell you the exhaust gas temperature at each engine speed as the engine

Carb Adjustments & Related Hardware

This is a 1971 Thermo-Quad carburetor. It was only used in '71 and has been superseded by the '72 Thermo-Quad. New parts for these '71 units are very hard to find and it is best to swap on a '72 and newer unit.

You want to use an actual fuel pressure gauge to determine your fuel pressure and it should be installed close to the carburetor.

accelerates to the peak power rpm. Based on the answers to this test, you can add "dams" to your manifold to solve the distribution problems. Dams are rarely added to dual-plane manifolds but are common on single-plane manifolds. Dams are typically made of wood that is epoxied into place. The wood dam is thin and maybe about 1/2" tall (on edge) so they are often cut to length from a popsicle stick. Popsicle sticks only work if someone has run your intake manifold model on a dyno, and has tested the size, location and number that define the details for the layout of the dam. Do not just add dams to any manifold without the dyno data, which you might be able to find from the manifold manufacturer or by searching on the Internet.

Usually one or two dams get the job done, but there have been cases that required as many as four dams. Oddly enough, an eight-barrel inline intake does not generally require dams but this may be an oversimplification.

Oil Shield

Typically on a V8 engine, the intake manifold covers the engine's tappet chamber. This means that the bottom of the intake manifold will be exposed to hot oil. This hot oil will make the intake charge warmer and this isn't good for overall power output. Some engines solve this by adding a shield below the manifold to keep the oil away from the manifold runners. Edelbrock offers a line of Air Gap manifolds for most V8 engines. The bottom of the Air Gap design acts as the shield for keeping the hot oil away from the bottom of the intake runners.

Carb Swaps

The production Carter carbs—AFB, AVS and Thermo-Quad—haven't been available new for more than 20 years, so if you want or need a new carburetor you will have to go with the Edelbrock.

The wide variety of service and tune-up parts that are available for the Edelbrock carburetors also make them a good choice. The '71 Thermo-Quad was a one-year wonder and should be swapped for the '72–'73 versions. If you have a Thermo-Quad engine and manifold and need a new carburetor, you might want to select an Edelbrock AVS. It fits on the Thermo-Quad style intake manifold with an adapter available from Edelbrock.

Fuel Delivery

The fuel delivery system should include a fuel filter after the pump and a fuel pressure gauge that should be located near the carburetor. The needle and seat assemblies are designed to work with 5 to 6 psi fuel pressure at idle. This will drop at higher rpm. If you use electric fuel pumps in the rear, next to the gas tank, and remove the mechanical pump, then you will need to add a fuel pressure regulator and set the fuel pressure to 5–6 psi.

Note: For street or cruising use, especially in hot weather, add a vapor separator filter right after the mechanical fuel pump. The filter has two outlets—one returns the vapor to the gas tank or to a charcoal canister, which then cycles it back through to the engine. These filters were stock on '70–'71 426 Hemi and 440 six-barrel production packages. To make them functional you have to add some fuel line but it is well worth it if you cruise in hot weather.

You also want a fuel filter in your fuel line but if you are planning on cruising in the summertime, it is a good idea to use a vapor separator like this unit.

Production heads can be helped to increase the airflow and make more power but you can also swap for newer heads, either aluminum or cast iron, that flow a lot more air and potentially can make more power. These gains must be included when you calculate your calibration package.

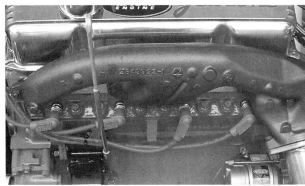

The exhaust system is very important to the carburetor. Most engines have high-performance cast iron exhaust manifolds available like these and headers (p. 133).

A small overbore on the block is not too big of a change but changing the stroke of the crankshaft by 0.4" or 0.5" is a big change in displacement and must be included in the carburetor's calibration calculation.

Most production exhaust manifold sets have a heat-cross-over valve like this that is used to make exhaust gas pass through the heat crossover passage under the production intake manifold that helps the engine warm up faster and have been fuel distribution which is related to drivability. If the heat crossover is blocked or removed for racing then the heat valve should also be removed or held open manually and the carburetor adjusted for the colder temperatures.

INDIRECT HARDWARE FACTORS

Virtually every engine part relates to the carburetor through the A/F ratio. Some of these relationships are somewhat obscure, so I will leave those to your imagination. In the following sections, I will cover some of the better known parts, like cylinder heads, cranks, headers, and cams and how these more obvious parts can affect carburetor adjustments.

Cylinder Heads

In the production Carter carburetor days, there usually were only one or two choices for cylinder heads. Today, almost all V8 engines have too many choices to name (cast iron, aluminum, etc.) especially the popular ones. Airflow increases are what you want to compare. If you flow more air, the engine tends to make more power and therefore the carb should deliver more fuel. Remember that the airflow gain may not be just due to the cylinder head but the ports (cnc-porting), valve size and shape and valve job.

Flow Bench

If the cylinder heads are going to be modified, then a flow bench must be involved. Once the head has been flowed on the flow bench, you should have a curve, not just a peak flow number. You should also have both sides flowed—intake and exhaust—not just the intake side that generates the higher/bigger number.

Displacement

Typically a production block is limited to an overbore of 0.020" to 0.060" but today special blocks are readily available that allow much bigger bores. The small overbore in an engine rebuild doesn't usually change displacement enough to affect the A/F ratio relationship with the carburetor.

Carb Adjustments & Related Hardware

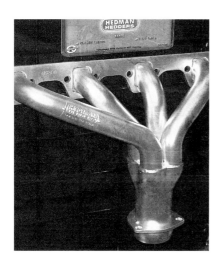

One approach to exhaust systems is to replace the cast-iron exhaust manifolds with a shorty header like this unit. These are popular on street rods because of limited space.

There are many mufflers that can offer less exhaust backpressure while controlling the noise level like this one from Flowmaster.

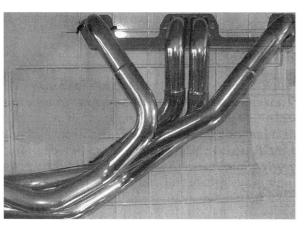

Smooth, high-flow headers will make more power. Street packages like to use Tri-Y headers while race cars use 4-into-1 like these. The 4-into-1 design generally makes more power and is tuned to the specific engine and application. With race headers the designer tries to make the bends smooth and the primary lengths equal.

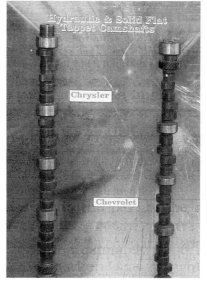

Cams are one of the most popular performance parts and they can greatly affect the carburetor. The most important aspects of cams for the carb are the duration and the overlap, which is directly related to the cam's centerline.

The cam manufacturer grinds the cam on a specific cam centerline, which can't be changed without a new grind. This ground centerline determines the cam's overlap. When you install the cam into your engine, the centerline between your crank and the cam can be adjusted or changed. Usually this is done by using offset keys or offset bushings, but it can also be done by changing the chain and sprockets or adding a belt-drive like this one.

Today there are many big-bore blocks, along with long-stroke cranks, both of which can greatly affect displacement. Remember, more cubic inches means more fuel will be required.

Exhaust

Production engines typically came with cast-iron exhaust manifolds. Standard engines came with low-performance, log-style manifolds, while performance engines used a more streamlined design. The aftermarket makes headers that are exhaust manifolds made from steel tubing and they come in many styles or designs—Tri-Y, 4-to-1 race, shorty, adjustable, step and many more. Generally headers are designed to make more power from the engine.

The second part of the exhaust picture is the mufflers. Production mufflers are quiet but somewhat restrictive but performance packages offer upgrades in mufflers. The aftermarket offers some very low restriction mufflers based on some unique technology. On most of the newer cars, there is a catalytic converter, which is related to the emissions package used on the engine. A catalytic converter also contributes to the exhaust backpressure. There are aftermarket companies that make larger diameter exhaust pipes and tailpipes and some performance models use an H-pipe to tie the left and right sides together under the transmission.

The carburetor tends to look at the net backpressure and this is related to all of those items listed earlier—manifolds, mufflers, catalytic

No matter which cam drive system you use, you should always check your cam centerline as you install the cam to be sure that it isn't off. This centerline check requires a degree wheel (shown) and a dial indicator, along with a pointer, which can be made from a coat hanger.

The top of the piston can tell you a lot. This is a domed piston but not too high so it is probably for an 11-to-1 compression ratio or similar. This is important because 10- and 11-to-1 engines require a higher octane than pump gas and the carburetor cannot be adjusted to solve this problem.

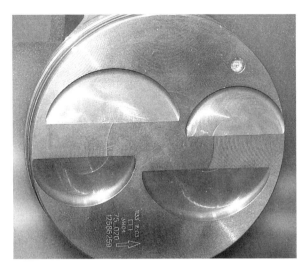

If your piston top has either 2 or 4 valve notches, then it is probably a 10-to-1 design and you should plan on using better gas.

converter and exhaust pipes. Less backpressure means more fuel.

Cams

Cams are one of the most popular items to change in the engine. The options are almost unlimited. Cams come in many sizes defined by lift, duration, centerlines, overlap and installation aspects. They are most often discussed and probably selected by lift and duration. The key for the carburetor is the amount of overlap. High cam overlaps push most of the fuel out the tailpipe so the carburetor has to compensate for this situation. More lift and duration generally means more power and higher engine rpm, and this requires adjustments also. Street engines want cams with wide cam centerlines (116–118 degrees) to keep the overlap to a minimum, which greatly helps drivability and therefore the carburetor.

Pistons

The piston itself is not the cause of the problem but it is most often the easiest solution. The problem is compression ratio, and indirectly the fuel quality. The fuel for any street car realistically is going to be pump premium—92 octane. This means that your engine should have a maximum compression ratio of 9.0-to-1. The resulting mechanical compression ratio for the engine is a result of the piston, the combustion chamber in the head and the block's deck height and cylinder head gasket selection. Typically by the time you find out that you have 10-to-1 or 11-to-1 compression ratio, it is too late to fix most of the pieces, so the piston is the part that can fix it—drop the compression ratio back to the desired number. The number is higher for aluminum heads because the aluminum conducts a lot of heat away from the chamber and this complicates the issue.

More compression ratio can make more power and require more fuel but that is not the main concern. The main concern is detonation. As the compression ratio increases over the maximum for the fuel, detonation begins and once it becomes audible, you may think that the engine is lean and try to solve the detonation by adjusting the carburetor. There are many manufacturers that offer a wide range of piston styles and features like CP Pistons, JE Pistons, KB Performance Pistons, Diamond, Arias, Ross, Wiseco and Silv-O-Lite.

CR Pistons—Judging the compression ratio of a piston is not easy and not very accurate but a very general guideline can be of some help. Typically, if the top of the piston has a dome on it, then it should have an 11-to-1 or higher compression ratio. If the top of the piston is flat with 2 or 4 valve notches, then it is probably a 10-to-1 piston. A piston that has a dish in the center of the top is a low ratio piston so it probably has an 8-to-1 or 9-to-1 compression ratio. A fully flat piston top lands somewhere in-between at 9- or 10-to-1.

Ignition

Carburetors have been used with both point ignitions and electronic ignitions. Troubleshooting the engine can point at both the ignition and the carburetor. They are very closely related in observed characteristics. For example, an engine miss could be caused by the carburetor or the ignition. I'll

Carb Adjustments & Related Hardware

The old-style points and condenser package was commonly used in production up through the early '70s. Most of the cars/engines built in the muscle car era used a version of this ignition. I would strongly recommend replacing any point system with an electronic system because they are readily available, cost less and once set up, require very little maintenance.

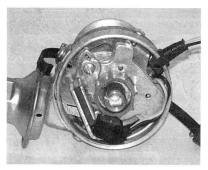

This is an electronic distributor, which you are required to use with electronic ignition boxes. Once set up, there is very little maintenance. This is a vacuum advance unit which is more desirable for street applications.

While most electronic ignition boxes are based on analog technology, there are some digital ignition systems that are available today. They use the typical electronic distributor shown at left.

In the front or in the trunk, the battery is a key player in the ignition's performance. If the ignition has problems, they may be mistakenly blamed on the carburetor.

The street and drag racing offer two different challenges to the cooling system. There are a lot of performance water pumps like this one-piece, billet electric water pump that bolts directly to the block. The electric water pumps offer performance gains for drag racing but haven't been tested for the various street uses.

leave that for the next chapter. For this section, you set the engine's total spark advance, which is based on the specific engine design but commonly around 35–38 degrees. For performance applications over the last 40 years or more, once you set the ignition's total advance, you install a fast advance curve in the distributor or replace the stock distributor with an aftermarket performance unit that already has a fast advance curve built-in. This fast advance curve means full centrifugal advance by about 2000 rpm. Over the last few years, we are pushing the fuel (pump premium) and the compression ratio (9-to-1) and the combination detonates in high-load, low-speed conditions. This translates to 2000–3000 rpm, and high gear. So today, you want to re-curve the distributor to drop the amount of advance, or delay it, at 2000–3000 rpm rather than be fully advanced at 2000 rpm. The detonation at 2500 rpm will sound lean but rejetting the carburetor doesn't fix it.

Note: In the next chapter, I'll discuss troubleshooting and calibration driving in the 2000–3000 rpm range is important to these steps and having detonation in this area will confuse the issue, perhaps point you in the wrong direction.

Note: From the carburetor's standpoint, there really is nothing wrong with points. The real problem is that they wear quickly and lose performance quickly also. That means to maintain the performance characteristics you have to constantly adjust the points, which customers aren't going to do. It is easy to try to adjust the carburetor to solve idle and drivability concerns. The solution to these problems is an electronic ignition which is almost maintenance-free and readily available. Performance distributors and electronic hardware like coils are available from manufacturers like Accel, MSD and Mallory.

Battery

The car's battery is very important in the performance equation. Most new batteries should hold about 12–14 volts. If the battery gets down in voltage, it can start the engine but not run it properly. The carburetor may be mistakenly blamed for the resulting misfires. This situation is even more likely if the battery is moved into the trunk.

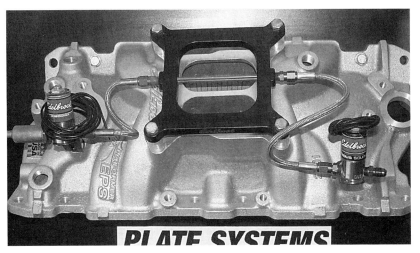

Nitrous systems are popular add-ons for increased performance. My recommendation is to select one manufacturer and purchase his compete nitrous kit.

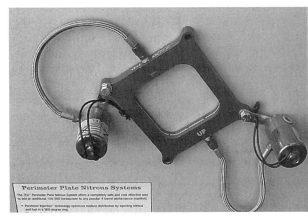

There are many styles of nitrous systems. This one uses a spacer plate and injects the fuel and nitrous around the perimeter. There are two solenoids—one for fuel (gasoline) and one for nitrous.

Many nitrous systems are very complicated. This one has four solenoids and several hoses that are all independent of the fuel line for the carburetor itself.

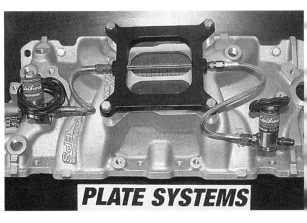

Another style of plate using a bar that passes straight through the spacer plate and injects the fuel/nitrous.

Water Pump & Cooling System

The engine's water pump is the heart of the cooling system. Generally the production cooling system is fine. However, you have increased the displacement, increased the horsepower and the cooling system hasn't been upgraded. Add to this the consideration that today you might want to cruise in the car—low speed, generally hot weather and long durations. If the cooling system can't keep up, the engine will tend to overheat, may go into detonation, or both. You may assume that the engine is lean, and try to correct the problem by adjusting the carburetor. Racing water pumps are available from manufacturers like Meziere and CSR Performance Products.

NITROUS OXIDE & CARBS

Nitrous systems come in all sizes and shapes. There are hidden systems, plate systems, and direct-port injection systems, multiple injectors, and combinations of all three. While the advertised horsepower gains are easily over 500 horsepower for some of the race systems, there seems to be fuel limits. If you are using only pump gas (92 octane) or pump gas plus octane booster, then the horsepower gains come down—100 hp to 200 hp appear to be realistic (remember our street focus). This might be considered the first-step nitrous system. Using pump gas, these street systems require less spark advance by 2–4 degrees total advance.

Note: Baseline engine is the specs for naturally aspirated high-performance engine. In the race-only

Carb Adjustments & Related Hardware

Either of the plate-style nitrous systems offer good performance gains and are easy bolt-on systems.

The blower drive belt and the cog pulley that drives the supercharger off the crank is to the right. The supercharger itself sits on top of the intake manifold and between the valve covers and takes up a lot of space. Because the carburetor(s) sits on top of the blower, plan on losing lots of hood clearance.

With a supercharger system, the supercharger itself goes between the carburetor(s) and the intake manifold. Typically the supercharger is driven by a large belt off the nose of the crank.

The street blowers take up somewhat less space and the drive belts are smaller/thinner but they still offer great performance gains.

systems, this spark advance decrease goes up to as high as 12 degrees. Additionally these street nitrous kits require colder spark plugs than the standard engine by 1-2 steps. For example, if you are currently using an 11-heat range plug (Champion heat range numbers), then you would drop down to a 9-heat range plug with a street nitrous kit. In the race-only kits, this cold plug requirement goes up even more.

Note: I'll leave the multi-steps/stages to the pros and nitrous manufacturers.

With a nitrous system, you inject both nitrous and fuel at the same time. The direct-port injection, which requires eight nozzles, injects both the nitrous and fuel through the same nozzle. The second style of nitrous system uses a central plate. Most off-the-shelf kits are based on four-barrel carburetors. A nitrous system has a lot of parts that are completely separate from the engine hardware. Solenoids and nozzles, and hoses and fittings and switches, plus the large nitrous bottle itself. I would strongly recommend getting a complete kit from one selected manufacturer like ZEX or Nitrous Supply, Nitrous Express or Nitrous Oxide Systems.

SUPERCHARGERS & CARBS

There are many ways to look at a supercharger system. Typically, the supercharger is adapted to the intake manifold and then carburetors are added to the top of the supercharger. While dual four-barrels tend to be the most popular system, the four-barrel carbs could be adapted in the same manner. On a plenum system typical of one of these supercharger systems, the eight-barrel system could be rated at about 1000 cfm, 1200 cfm or 1600 cfm.

There are many racing-based superchargers and perhaps even more street blowers. The older-style racing-based blowers are called 6-71 and 8-71 (GMC) blowers along with the similar aftermarket designs (clones). They sit on top of the engine's intake manifold and are driven by a cog belt off the

Similar to the supercharger, you would mount the turbocharger unit between the carburetor and the manifold/cylinder head. Since the turbo is driven by exhaust gases, this isn't as easy to do as it is on a supercharger. On the other hand, you don't have to consider the large drive belt being attached to the nose of the crank.

front of the crank. There are many street blowers that are belt-driven. There are two main issues when you are trying to adapt any of the blowers or superchargers to the general four-barrel engine—compression ratio and boost. Since you are building a street car, then you will run pump gas and that has a limited octane (92 max), which limits the supercharged engine's compression ratio to about 8-to-1 max. The second issue is limiting the boost to about 6.5 to 8 psi max. Remember that the AFB/AVS carburetors with metering rods allow you to fine tune the engine's calibration without taking the carbs apart.

With a supercharger system, most of the manufacturers try to match the supercharger size to the engine's size—displacement. The typical supercharger size is also listed in cubic inches of displacement. You would probably not want to use the same size supercharger on the 350 and 500 cubic-inch engines. A 350 or 375 cubic-inch engine might use what is rated as a 112 blower while the 450 cubic-inch and up engines might want a 122 blower. Another issue with these street or dual-purpose superchargers is the length of the typical carb pad opening. Some of the more popular supercharger manufacturers are Magnuson Products, Whipple Industries, and Eaton Performance Products. Blower drives for various engines are available from manufacturers like The Blower Shop and BDS—Blower Drive Service.

Tip: Edelbrock recommends using two #1405 carburetors for blown engines. They also recommend the following calibration—primary jets, 0.101"; secondary jets, 0.101"; metering rods, 0.070" x 0.042"; step-up springs, 5" (orange); needle and seat assemblies, 0.110". They recommend this calibration for 350 to 440 cubic-inch engines.

TURBOCHARGING & CARBS

A turbocharger system is somewhat like a supercharger except that the exhaust drives the turbocharger where a belt drives the supercharger. With turbo kits, it is very important to use lower compression ratios like 7.5 or certainly below 8.0. A similar situation exists with boost pressure with limits of 8–10 psi. This is all related to the pump gasoline being used. One carburetor would work well with one turbo but if two turbos are used on a V8, then you should consider two carburetors. Turbos are available from Turbonetics and Honeywell Turbo Technology.

Chapter 7
Carb Troubleshooting, Tune-Up & Calibrations

Before you begin any carburetor calibration, install new spark plugs.

Perhaps "tune-up" is the most general term because it would cover fixing, repairing, rebuilding and improving. With the term "diagnosis" you tend to think of defining a problem. With troubleshooting, you tend to think of locating a problem. You may think that you have trouble but by defining the problem and locating the problem, you will have a much better chance of solving the problem. With calibration or tuning, you may think of a spec sheet, formula or recipe that will allow you to get from defining and locating to solving. What I will try to explain in this chapter is all of these. The end result or goal is to improve the overall performance of your package.

ENGINE DIAGNOSIS

After the carburetor has been installed onto the engine and the engine has been started, you will want to tune it up. There are many ways to start this process but you need a diagnosis of the engine. The main problem that you encounter in doing a diagnosis is that many of the trouble areas, or symptoms, are the same for the fuel-related items and ignition system items. The characteristics are identical, in many cases.

Ignition

For example, an engine miss can be caused by fuel calibration problems or spark and ignition troubles. While old spark plugs are an obvious potential problem, do not forget the spark plug wires. This is especially important on older engines that still have the original plug wires or wire

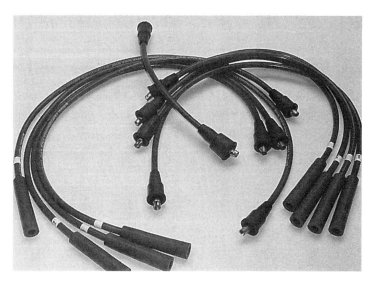

On any engine project, you should replace the original spark plug wires. Use a new set with the cylinder numbers marked on the ends to make it easier to connect them with the proper cylinder. You would also like to have them premade because cutting wires to length means you have to finish one end by hand, which is difficult to do. Also for best engine performance, you should keep the coil wire as short as possible. If you have a long coil wire, move the coil closer to the distributor.

REBUILD & POWERTUNE CARTER/EDELBROCK CARBURETORS

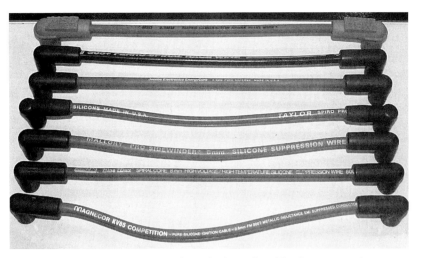

There are many different styles of spark plug wires. The boots or ends vary somewhat. Production plug wires tend to be 7mm and many performance wires are 7.5mm. For racing there are 8mm and even 9mm wires but these aren't required for street or dual-purpose applications.

It is much easier to do any troubleshooting, diagnostics, or carburetor calibration with a distributor and ignition that doesn't require constant maintenance, so I'd strongly recommend replacing any point ignition with an electronic unit.

For any troubleshooting or calibration work, you will need a good tachometer. You also want to mount the tach in a place that is easy for you to see.

sets that have been reused for many years and many sets of spark plugs.

Battery Voltage—In many cases it is easier to check the ignition aspects than it is to change parts of the carburetor. One of the first things to check is the battery voltage. It should be 12 volts or higher. The engine will miss at around 11.8 volts (typical of electronic ignitions), but that 11.8 voltage is still adequate to start the engine in most cases. If in doubt, replace the battery with a new one.

Plug Wires—Next are the spark plug wires. If they are old, you should replace them. Pay particular attention to the coil wire. Old plug wires can easily cause a miss that can't be fixed by recalibrating the carburetor. Next are the spark plugs—did you forget and leave in the old plugs that you used to cover the holes during the engine installation?

Vacuum Leaks

Vacuum leaks will make the engine run lean. Replace old vacuum hoses because they tend to dry out and crack and cause vacuum leaks. You may need a spray bottle with soap and water to go over the engine if you suspect a vacuum leak.

Diagnostic Tests

Once you have the engine started and broken in, you can run some tests on it. Before you get into troubleshooting the induction system, you would like to make sure that any observed problem isn't caused by something in the engine itself. It is much easier to check on the various aspects of the engine than it is to do the same thing for the induction system. In the process of warming up the engine and breaking in the cam and rings, you can usually check for vacuum leaks and double check the plug wires and battery voltage discussed earlier in the chapter.

Timing—During the engine build, the distributor is installed as close to TDC as desired and once the engine is started, the initial timing on the distributor is reset to the number desired. This should be done before the carburetor idle is set. The total timing is also very important to overall engine performance. The centrifugal advance is built into the distributor. The total advance is the sum of the initial advance added to the centrifugal advance. If too much total advance is set, there can be detonation issues.

Note: Assume a centrifugal advance of 30 degrees and an initial setting of 8 degrees before—then the total advance is set at 38 degrees. If you only wanted 35 degrees total, then the initial setting has to be lowered to 5 degrees.

Carb Troubleshooting, Tune-Up & Calibrations

It is very important—before you get too far into any final calibration or troubleshooting process—to check the cam installed centerline. You need a degree wheel, as shown, and a dial indicator along with a pointer that can be made from an old coat hanger.

Fuel Pressure—It is almost as important not to have too much fuel pressure as it is to not have too little fuel pressure. Or too much is as bad as too little. Generally carburetors want a fuel pressure of between 5 to 7 psi. To determine this, you need to have a fuel pressure gauge. Mount it where you can read it easily, close to the carb.

Centerline Camshaft—When the engine is built and the camshaft is installed, the cam's centerline should be installed using a dial indicator and degree wheel. It can be installed by lining up the dots, but it is very easy to end up with a cam installed a tooth off and even this amount would greatly affect the calibration tests. The camshaft also has a centerline that is ground into the camshaft by the manufacturer but it can't be changed once the cam is made. For street engines, you would like to have a wide cam centerline—114 to 118 degrees.

Fuel Quality vs. CR—Assuming that the fuel to be used is pump premium or 92 octane, the engine's compression ratio should be 9.0-to-1 or less (10.25-to-1 for aluminum heads). If the compression ratio is higher than this, you will have detonation problems, especially after the first 1,000 miles.

Compression Test—Once the engine has been broken in, you can run a compression test. Even if nothing is wrong with the engine and you are not looking for problems, it is still a good idea to run tests when it is new and in good shape. This information can be handy later on when you think something has gone wrong in the engine. A compression test is easy to perform (see sidebar above).

Leakdown Test—The leakdown test is similar to the compression test except that it tends to offer more information. However, the leak test requires a special tool (leak tester) and compressed air source (around 100 psi). Similar to a compression test, do not consider the absolute number(s) as much as any variation from one cylinder to another. A leak test number(s) can give more of a quantified answer but the operator of the device/test should give you comparative numbers based on previous tests he has run with that specific tester. See sidebar on page 142.

COMPRESSION TEST

The following is a method to test the engine's compression. If something has happened to the engine's sealing capability, such as failed piston rings, a blown head gasket, burned or bent valves, then one or two of the cylinders will have very low compression readings. This test assumes that the battery is fully charged. More or less mechanical compression ratio built into the engine will affect these very general numbers.

Tip: It is helpful to keep a battery charger hooked up to the battery during the test to ensure that the cranking speed for each cylinder is maintained.

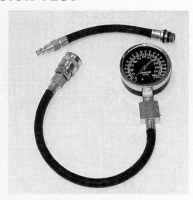

The compression gauge usually comes in two parts—the gauge itself is mounted on a section of hose with the quick-disconnect fitting. The other piece screws into the cylinder, seals it and connects to the quick-disconnect fitting.

1. Run engine to temperature and then shut off
2. Pull center coil wire (so engine will not start)
3. Remove air cleaner
4. Remove all spark plugs
5. Open throttle plates
6. Insert compression tester in the #1 spark plug hole
7. Crank the engine on starter—at least 4 cycles
8. Write down maximum reading on gauge for cylinder #1
9. Repeat procedure for all remaining cylinders

Note 1: Compression readings should be uniform in all cylinders with less than 25 lbs. variation, from best to worst.
Note 2: The absolute number is not as important as any cylinders way below the average.
Note 3: Obviously the speed (rpm) at which the starter turns the engine over factors into the compression reading.

An approximate compression pressure with the engine warm, plugs removed and throttles fully open.

Engine (cid)	Pressure (psi)	Pressure Variation (psi)
200–250	125	20
251–300	135	20
301–400	140	25
401–450	150	25

LEAK TESTING

A leakdown test is more accurate than the compression test. You will need a leakdown test gauge, available from Sun Electric or Snap-On. Most engine shops also have one. The leakdown test was developed as a racing/engine performance tool in the early years of the Hemi engine as a method of determining the basic capabilities and problems (if any) because there were so many engines of the same kind being raced across the country. To use, you need a compressed air system putting out 90 to 100 psi. The nice thing about a leakdown test is that it tells you where the leakage is actually occurring in the engine.

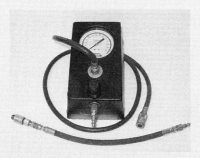

The leakdown test gauge is about the size of a small shoebox. It is similar to the compression tester in that it has a piece that threads into the cylinder head (bottom of photo) and another hose connected to the tester that has a quick-disconnect fitting on the other end. The major difference is that the leak tester has another fitting (in the center of gauge) that connects to a 100 psi shop air system.

To run a leakdown test, remove all spark plugs and rotate the engine to exactly TDC on the number 1 firing position (both valves closed). Screw the tester fitting into the spark plug hole. To gain the maximum amount of information, you need to have access to the radiator cap, or water jacket (water pump outlets), the carb air cleaner (removed), the headers and the crankcase filler cap. You will watch for or listen for air escaping (leaking). Once set up, hook up the machine. There is now 100 psi air in the cylinder and both valves should be closed.

- Air through the carb indicates poorly seated or bent intake valve.
- Air through tailpipe/header indicates bad, bent or burned exhaust valve.
- Air bubbling into the radiator indicates a blown head gasket or air escaping through water pump outlets.
- Air in the crankcase (valve cover) indicates bad rings/ring seal.

An engine always leaks a little! One of the nice features of a leakdown tester is that it gives you a numerical answer for how bad the leak is—3%, 10%, 20%, etc. And like the compression test, all cylinders should be close to each other in leakage. It could be a broken ring or something stuck, so don't panic—sometimes things stick and get better once the engine is run a bit and broken in.

A/F Ratio

The air/fuel ratio for the engine is generally discussed as being either rich or lean. Much of the following troubleshooting chart is based on rich-lean conditions in the engine. The A/F ratio is adjustable by using the carburetor's jets and metering rods. If there are basic troubleshooting issues, then they should be completed first. Once you have arrived at adjusting the A/F ratio, then I have outlined a procedure in the calibration section, following the troubleshooting section.

Quick Test for Rich or Lean—If you have a question concerning the engine's A/F ratio, whether it is too rich ot too lean, there is a quick test that you can perform. Drive the car with the air cleaner installed, and then remove the whole air cleaner assembly. Removing the air cleaner will lean the engine out or installing the air cleaner will richen it up. Whichever configuration works the best tells you which way to go, in general.

GENERAL CARBURETOR TROUBLESHOOTING

Once you have started the engine, warmed it up for the first time and done the engine break-in procedure, you have probably already completed the initial timing, vacuum leaks, and idle set procedures, so the following chart is a listing of the common areas to begin troubleshooting. If you have no problems other than items that you consider related to the A/F ratio (lean or rich) then you can skip to the calibration section beginning on page 146.

Try to consider the engine's performance by breaking the full range down into three sections—idle, part throttle and wide-open throttle. While not specific, I would suggest thinking of part throttle as engine speeds from 1800 rpm to 3000 rpm. Remember to adjust the fuel pressure to numbers acceptable for the carburetor being used. Also, before you start any calibration development, you should try to eliminate any concerns relative to detonation by using the best gas possible. If the car has been sitting for a long time while awaiting its new engine, you might want to replace the fuel that was left in the tank. Also check the in-tank filter on cars that have them.

CARB TROUBLESHOOTING, TUNE-UP & CALIBRATIONS

TROUBLESHOOTING CARBURETORS

Common carburetor troubles and possible solutions—assuming engine is OK mechanically

Trouble	Potential Cause(s)	Possible Solution(s)
Flooding or leaky carb prior to start-up	Cracked carb body Faulty body gasket High float level Worn needle and seat Excessive fuel pump pressure Leaking carb float	Replace carb body Replace defective gaskets Test and adjust float height Replace needle and seat Test pump pressure Replace float
Carb floods after start-up	Float level off Dirt/metal in needle/seat Filter plugged Air horn gasket not sealing Fuel pressure too high	Set float level and drop Clean/replace needle/seat Replace filter Replace air horn gasket Adjust fuel pressure
Hard starting (engine cold) choke plate stuck	Incorrect choke adjustment Binding choke linkage Choke plate stuck/jammed Choke thermostat corroded Air cleaner interferes with choke Fuel filter clogged Air cleaner gasket interferes with choke Improper fuel level in carb Engine backfires	Adjust choke Repair defective parts Unstick plate and free up/lube Replace thermostat assembly Remove interference Replace fuel filter Reinstall gasket properly Adjust float level Set engine timing
Low engine output (10°F or lower)	Engine oil—wrong viscosity Valve lash incorrect Choke thermostat adjustment off, running rich Timing set incorrectly	Recommended 5W-20 Set valve lash Adjust to correct setting Set timing correctly
Stalling (engine cold)	Choke system lean Choke vacuum diaphragm adjustment lean Engine idle speed too slow Intake manifold or carb gaskets leaking Defective fuel pump leaking	Adjust choke to close fully Adjust to specs Adjust idle up to spec Replace gaskets Replace fuel pump
Low engine output	Fast idle speed low Fast idle cam adjustment off Engine oil of incorrect viscosity	Adjust to specs Adjust to specs Recommend 5W-20 or 10-30
Carb—lean	Curb idle set very lean Air leak bypassing carb	Adjust idle to richen Repair vacuum leak(s)
Stalling (engine hot)	Improperly adjusted dashpot Idle speed too slow Incorrect idle fuel mixture	Adjust, repair or replace Adjust idle speed Adjust idle fuel mixture

(continued on next page)

(Troubleshooting Carburetors, continued)

Trouble	Potential Cause(s)	Possible Solution(s)
	Worn/bent idle fuel mixture screw(s)	Replace mixture screws
	Defective fuel pump	Replace fuel pump
	Coolant thermostat defective	Replace coolant thermostat
	Fuel tank vent clogged	Remove restriction or replace gas cap
	Fuel lines clogged	Remove restriction, tighten fittings
	Fuel filter clogged	Replace filter
	Throttle shaft too loose in body	Repair or replace parts as needed
	Incorrect throttle linkage	Adjust throttle linkage
	Leaking intake manifold gasket	Tighten bolts or replace gaskets
Stalling (on hard braking)	Fuel pressure too high	Set proper fuel pressure
	Floats have leaked	Replace floats
Runs excessively rich after cold start	Choke thermostat adjustment is rich	Readjust to correct spec
	Choke thermostat damaged	Replace thermostat assembly
	Choke vacuum diaphragm misadjusted	Correct or replace
	Choke vacuum passage blocked	Correct
	Incorrect gasket between carb and intake	Correct or replace
Engine won't idle	Air/vacuum leaks	Fix leaks
	Ignition system problem	Check and replace parts as needed
	Choke misadjusted	Adjust choke
	Idle mixture off	Set idle mixture screws
Rough idle (poor idling)	Incorrect idle adjustment	Perform idle speed adjustment
	Surging Idle (Neutral vs In-Gear)	Richen idle
	Damaged tip on idle mixture screws	Replace mixture screws
	Clogged idle passages	Clean carb body and blow dry
	Incorrect throttle stop adjustment	Adjust throttle stop screw
	Incorrect fuel level (float)	Adjust fuel level
	Incorrect fast idle cam adjustment	Perform all idle adjustments
	Choke misadjusted	Adjust choke
	Air cleaner restricted	Clean and replace element
	Worn throttle shafts	Inspect and replace carb body
	Loose main body screws	Tighten body screws securely
	Incorrect valve lash	Adjust valves
	Air leaks in intake manifold	Replace intake gaskets
	Fuel pump defective	Replace fuel pump
	Incorrect idle fuel mixture	Adjust idle mixture screws
Severe auto trans engagement (after cold start)	Fast idle speed set too high	Adjust fast-idle speed
	Binding or sticking throttle linkage	Repair defective parts
	Too lean idle setting (jets, metering rods, mixture screws)	Richen idle setting
Poor low-speed operation	Idle adjusting screw incorrectly adjusted	Perform idle speed adjustment
	Clogged idle transfer holes	Clean carb body and blowdry
	Restricted idle air bleeds	all passages

Carb Troubleshooting, Tune-Up & Calibrations

Trouble	Potential Cause(s)	Possible Solution(s)
Poor acceleration	Accelerator pump piston not sealing	Replace accelerator pump
	Faulty acc-pump discharge ball	Replace discharge ball
	Faulty acc-pump inlet check ball	Replace inlet check ball
	Incorrect fuel or float level	Test and adjust float level
	Worn acc-pump and throttle linkage	Replace accelerator pump
	Fuel filter clogged	Replace fuel filter
	Defective fuel pump	Replace fuel pump
	Manifold heat valve sticking	Free up using solvent
	Incorrect pump setting	Reset pump
	Automatic choke malfunctioning	Repair or replace as needed
	Clogged vent in gas cap	Remove restriction or replace cap
	Clogged fuel line	Remove restriction
	Weights stuck in distributor	Fix distributor or replace
Surging at cruising speed or misses under load	Clogged main jets	Clean main jets and blow dry
	Undersize main jets (too lean)	Replace with larger main jets
	Low fuel level	Adjust fuel level (float)
	Low fuel pump pressure or vol.	Test fuel pump
	Blocked air bleeds	Clean carb body and blow dry
	Clogged inlet filter/screen	Replace fuel filter/clean screen
	Air/vacuum leaks	Check and replace gaskets
	Choke adjustment off	Adjust choke
Mixture too rich	Restricted air cleaner	Replace element
	Leaking float	Replace float
	High float level	Adjust float to spec
	Excessive fuel pump pressure	Test fuel pump
	Worn metering jet	Replace jet
Engine runs rich	Choke stays on	Adjust choke to pull off sooner
	Incorrect carb gasket	Replace with correct gasket
	Air cleaner restriction	Replace air cleaner
	Jets, metering rods too large	Install leaner rods/jets
Mixture too lean	Air leak bypassing carb	Repair vacuum leak
Engine misses under load	Mixture too lean	Install richer calibration
	Fuel pressure too low	Adjust fuel pressure
	Intake leak, manifold or gaskets	Repair leak
Engine runs lean	See hard starting, page 143	See hard starting, page 143
• first 1/2 mile—choke lean	Diaphragm adjustment lean	Readjust to specs
• after 1/2 mile when engine heat insufficient	Heat valve stuck open	Free with solvent
	Heat valve thermostat distorted	Replace thermostat
	Heat valve failed within exhaust	Replace heat valve
	Water temp subnormal	Check coolant thermostat
	No heat valve (headers)	Change part-throttle fuel mixture
Reduced top speed	Low fuel pump vol/press.	Test fuel pump
	Incorrect fuel level in carb	Adjust fuel level (float)

(continued on next page)

(Troubleshooting Carburetors, continued)

Trouble	Potential Cause(s)	Possible Solution(s)
	Main jets too small	Replace with larger main jets
	Faulty choke operation	Check choke for full-open position
	Clogged fuel filter	Replace filter
	Gas tank vent plugged	Fix gas tank vent
	Fuel line clogged or kinked	Blow out line or replace
	Air cleaner restricted	Clean and replace element
	Improper throttle linkage adjustment	Adjust for full opening
	Secondary throttle system defective	Repair
Secondaries fail to open	Carb linkage	Repair carb linkage
	Vehicle linkage	Repair and adjust vehicle linkage
	Secondary throttle plates sticking	Free with solvent and readjust
Low fuel economy	Choke stays partially on	Reset choke
	Mixture too rich	Install leaner calibration
	Ignition timing incorrect	Set proper timing
	Engine has detonation	Lower CR or use better fuel

Most initial calibrations are done with the air cleaner removed. Once you get close, you can install the air cleaner and use a long screwdriver to adjust the idle mixtures for use with the air cleaner.

PERFORMANCE TUNING

First step is to read Chapter 6 on the adjustments available. In Chapter 6, I tried to show the various adjustments that were available and why adjustments may have to be made. With so many options available for engine parts and specifications, you have to estimate your specific engine package compared to a baseline. Our estimates are not always perfect. In Chapter 6, I tried to use adjustments that could be done without taking the carburetor apart once it is together. In this section, I will try to keep that in mind, but if the calibration is off based on testing your specific package, then the installation of a new calibration can involve carburetor disassembly—Chapter 4.

Tools and Parts Needed

For basic calibrations designed to improve your performance (as opposed to fixing a problem), you'll need a tachometer and vacuum gauge and basic hand tools. If you do not have a vacuum gauge, you can just use the tachometer for the adjustments.

Assorted jets, metering rods and springs—Make sure you gather up the extra tune-up parts needed to calibrate the carb. These were covered in Chapter 6. Calibration kits are available from Edelbrock for AVS and AFB.

Idle Mixture Calibration

For this step the engine should be warmed up and the choke should be fully out (off)—blade straight up/down. The air cleaner should be on (installed).
1. Set the desired engine idle speed (rpm) using the main primary throttle idle screw.
2. Adjust one side idle mixture screw to obtain maximum rpm (or the max vacuum reading).
3. If the idle speed changed more than 40 rpm, readjust idle screw.
4. Repeat procedure on opposite side.
5. Reset idle speed.

WOT Calibration

It is best to do a wide-open throttle calibration on a chassis dyno, since they are quite common and you don't have to remove the engine from the car. If a chassis dyno session is not in your budget, then you can select a safe, legal driving space like a drag strip.

Carb Troubleshooting, Tune-Up & Calibrations

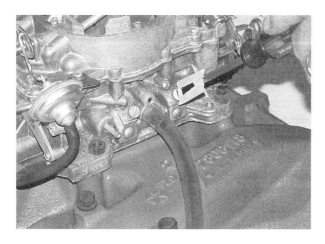

When you are adjusting the idle mixture screws on the front of the carburetor, you might want to place a small piece of masking tape on one side of the screwdriver and place a black mark on it to help you count turns.

For this test you will need a stopwatch and a notepad. First you have to select an rpm range for your test for evaluating the WOT power of your engine package. You should perform this test over about 50% of the powerband. If you have no specific dyno data, start with the engine's peak power rpm. Use the estimates provided in the chart in Chapter 3 on page 36.

Some camshaft manufacturers provide some guidelines that can also be helpful. For example, a street hydraulic cam might have a power peak of 5500 rpm. Dividing 5500 by 2 yields 2750, so your test would run from 2500 to 5500 rpm. If you had a mechanical cam, the peak speed for a street package might be 6500, so you could use 3500–6500 for your test range. Obviously gauges are very important for this test, and a wide range of automotive gauges are available from AutoMeter Products.

You should start your test below the minimum rpm range so the engine is accelerating smoothly through the test speed. This produces more consistent, accurate results. My recommendation for a starting speed on the hydraulic is 2000 rpm—500 below the test minimum. For higher minimums, you can use 500–1000 below the lower limit. If you do not have a chassis dyno or access to a drag strip, then consider what gear you want to run the test in. I would recommend second, because with a high-performance engine package, you may have too much wheelspin in first gear and the vehicle speeds can become too fast in high gear. Time your acceleration with a stopwatch.

1. Make 3 runs exactly the same and average the speeds for a baseline. Always test in the same direction so the wind and road grade, if any, are not factors. If, as in our example the range is from 2500 to 5500 rpm, start the test at 2000, but only beginning timing during that range.
2. Install a metering rod that is 2 stages or 8% richer—on small step or bottom step. Install the same rod on both sides.
3. Repeat acceleration runs and average the speeds.
4. Compare 8% richer times to the baseline.

If 8% Richer is Faster—If the 8% richer setting is faster, then you should change the secondary jets 2 stages richer and perform the test again. Compare acceleration times. If they are the same, you are done. But if the new richer calibration is slower, change to 1 stage rich for the primary and secondary jets (compared to baseline package) and you are done. If the new richer calibration is faster, then go 3 stages rich on the primary and secondary jets.

If 8% Richer is Slower—Now, if the 8% richer calibration is slower than the baseline calibration, install a 1-stage leaner primary jet and a 1-stage leaner secondary jet and repeat acceleration tests. If these results are the same, go back to the baseline setting. If these results are faster, go 2 stages lean on both primary and secondary jets.

If 8% Richer is the Same—If the two results are the same do not be surprised. Change back to the base calibration. This means that you did a good job of estimating the jet sizes and selecting the metering rods.

Part Throttle Calibration

There are many throttle positions between closed (idle) and wide open, so this is a tough one to define. Edelbrock uses two basic ones, Cruise Mode and Power Mode.

Cruise Mode—Is defined as a low power, steady cruise with light acceleration. This means that the intake manifold vacuum is high and the metering rods are down in the lean position.

Power Mode—Is defined as high power and heavy/high acceleration but not wide-open throttle. I might put this in as 3/4 throttle. In this configuration, the manifold vacuum will be low, and the metering rods will be up in the rich position. Remember that the rich position for the metering rod is the small or bottom end.

With either of these two modes there are no absolute performance numbers. You drive at various vehicle and engine speeds (rpm), and various throttle openings in this range and you try to sense flat spots or lean-surge conditions.

If you find a surge or a flat spot, then install a 1-

stage richer metering rod. This means that the small end of the metering rod stays the same and you change the cruise or part-throttle step on the metering rod to one step smaller (richer). Do both sides the same amount and at the same time.

Edelbrock has detailed calibration charts for each carburetor that they offer. For example, a part-throttle or cruise mode step on a typical metering rod is 0.065". Dropping this 1 step richer means it goes to 0.062" and this change equals a 4% change in the richer direction for the calibration.

For the power mode part of the part-throttle range, this probably means that the manifold vacuum is 5" Hg or less. Remember that the bottom of the vacuum chart—zero or close to zero—means that the throttles are wide open and I selected about 3/4 throttle for the upper end of the part throttle range, so you should be above the bottom of the curve. A lean drivability symptom (surge or flat-spot) in this power mode means 1 stage richer metering rod—part throttle step. You might also try a stronger step-up spring with the metering rod.

Metering Rod Step-Up Springs—The step-up spring installs under the metering rod piston. There are five different springs rated soft to strong. They are color-coded. Blue = 3" Hg. Yellow = 4" Hg. Orange = 5" Hg. Pink = 7" Hg. And Plain = 8" Hg. The middle spring or orange is the typical starting point spring. If you find a flat spot or surge in your calibration tuning, changing the step-up spring is one possible solution. If you find a flat-spot as you gradually open the throttle, you may need a stronger step-up spring—which means plain or pink.

Accelerator Pump Calibration

The accelerator pump should be left until after all the basic calibration work has been completed. The accelerator pump is designed to help the engine change speeds (rpm). So if you have a hesitation or stumble as you change engine speed that is not related to the basic metering calibration, then you may need more accelerator pump. At that point, move the pump drive link, the upper pivot, closer to the center pivot/carb body. This change will increase the stroke length of the plunger and result in more pump fuel delivery to the engine.

Float Adjustment

The float adjustment was done in Chapter 4. All the Edelbrock AFB and AVS carburetors use a float setting of 7/16" and a float drop of 15/16" to 1". Production AFB and AVS carburetors may use a slightly different specification. The exception are engines used in off-road applications (see page 149).

Choke Adjustment

There are electric chokes and manual chokes plus many versions of the production style chokes. The key is that the choke come on (close—almost horizontal) when the engine is cold and that it goes off fully—vertical (off)—after the engine is warm.

Fast Idle Cam

The fast idle cam is a separate part of the throttle linkage that allows the engine to operate at a slightly higher engine speed when the engine is cold. Typically the fast-idle speed is around 1500 rpm. Once the carburetor is installed onto the intake manifold, the fast idle adjusting screw is on the bottom of the linkage. Therefore shut off the engine when adjusting this linkage. Generally you have to open the throttle to move the screw out and up so that you can access it for adjusting.

Long Duration Cam

It is common for long duration cams to cause excessive throttle openings as the idle setup on the carburetor. This excessive throttle opening causes low-idle vacuum levels and leads to poor adjustability and erratic idle speeds. One of the problems with this situation is if the engine's vacuum is below 7" Hg., then the metering rods will be in the up position. The up position means that the small part of the metering rod is now in the jet and this position is rich. One solution is to install a weaker step-up spring and this will help bring the metering rods down at idle. The typical standard step-up spring is orange and weaker step-up springs are blue or yellow (weak).

Note: The stronger springs are plain or pink. Refer to the Metering Rod Step-Up Springs section for more detail on step-up springs.

One of the problems with long duration camshafts is their higher amount of overlap. Having the longer duration cam ground on wide centerlines will help this situation. If that is not an option, then the higher overlap means both valves are open longer and this allows more fuel to come in the intake valve and go right back out the exhaust valve into the exhaust manifolds. By the time the exhaust valve gets closed, most of the fuel has gone into the exhaust system, so what is left has to feed the whole cylinder, and the net result is that the engine runs very lean, especially at low rpm or idle. To solve this lean condition, the throttles are commonly adjusted open to get to the main metering system for more fuel. With the throttle blades open and lots of spark advance (fast advance curve in performance distributor) the engine tends to surge—it goes fast,

Carb Troubleshooting, Tune-Up & Calibrations

maybe 1500 to 2000 rpm and then it goes slow, maybe under 1000 rpm and keeps cycling back and forth. The easy test for this condition is the in-neutral and in-gear idle tes with auto transmission. If there is more than a few hundred rpm difference between these two numbers, then you need to richen up your idle.

The first step is the idle mixture screws. Turn them out (richer) by 1 turn. Remember that the production AVS uses a left-hand thread on the adjusting screw while the AFB, TQ and Edelbrock AVS use right-hand threads. If you get to about 3 turns out total from the seated position, then you should consider a richer metering rod—part throttle step. Remember that richer means smaller diameter. If that doesn't solve problem, then you will have to try a richer primary jet.

Note: The TQ has an additional metering rod adjustment on the center piston—see Chapter 6.

Off-Road Calibration

The term "off-road" has been used to mean almost any kind of racing but in this sense, I am referring to its somewhat original meaning of hill-climbing or desert racing. The Edelbrock AFB and AVS carburetors used in this style of off-road racing should use a spring-loaded needle-and-seat assembly—Edelbrock #1465—to replace the standard needle and seat. This type of off-road racing tends to put the vehicles in very high angles—both front-to-rear and side-to-side. The standard needle and seat are happy on basically flat terrain like a drag strip. The spring acts somewhat like a shock absorber and helps reduce the possibility of flooding during these off-road maneuvers. Once installed, reset the float height to 7/16" and the float drop to 15/16" to 1".

Note: You can also consider a vacuum fuel pressure regulator for off-road use.

High Altitude Calibration

Elevation affects the calibration in the carburetor. A common rule of thumb is that you want to allow for a 2% leaner condition per 1,500 feet of elevation change. So if you developed your calibration package at sea level (zero elevation) and then went to 6,000 feet elevation (Denver is at about 5,200 feet), then you need to change your engine's calibration package by 8% (there are four 1,500-foot steps in 6,000 feet, so 4 x 2 = 8%). Depending upon your specific package, this might be a jet and metering rod change, but if you made this trip regularly, then I would recommend setting it up so that it is simply a 2-step swap on the metering rods.

This is an MSD electronic box called the MSD 6, which is very popular in many applications today.

Blended Fuels

A blended fuel is a mixture of standard gasoline and alcohol. The most common is called Gasohol. It is typically made with ethanol. As long as the percentage of alcohol that is mixed with the gasoline is not more than 10%, you can use the standard gasoline calibrations discussed earlier. If the percentage is more than 10%, such as E85, which is 85% ethanol, then these blended fuels will require a richer A/F ratio—bigger jets.

Supercharged Engines

While there are many possible options for blown engines, Edelbrock has selected the most common or most popular package for which to make a specific recommendation. Edelbrock recommends two model #1405 AFB carburetors for use with a supercharger. The supercharger itself is a GMC 6-71 or equivalent. There are several other manufacturers that have similar sized, positive displacement, belt-driven superchargers.

Primary Jets: 0.101"
Secondary Jets: 0.101"
Metering Rods: 0.070" x 0.042"
Step-Up Springs: Orange
Needle and Seat Assemblies: 0.110" (standard is 0.0935")

This package works well for 340–450 cubic-inch wedge-head engines

IGNITION TESTS

I have discussed the initial timing setting and the total advance on page 140 of this chapter. In this section, I want to address the curve—how the distributor gets from initial reading to total spark advance. If a distributor has 30 degrees built in, that means that the distributor's centrifugal advance is 30 degrees and the distributor's advance curve is how it gets from zero degrees to 30 degrees.

Note: Typical initial settings are around 8 degrees and many distributors have a built-in centrifugal advance of 30 degrees. This means that the engine's total advance is 38 degrees. If you want 32 degrees total advance, then you set 2 degrees initial. The curve is how the distributor controls the 30 degrees—fast or slow, etc.

Many aftermarket distributors have different amounts of centrifugal advance, perhaps 20 degrees or 30 degrees. In some cases, the amount can be changed or adjusted. The key for today's engines being used on pump gas in cruising and street use is how it gets from zero to the full amount. For just performance/racing applications on race gas, you wanted a fast advance curve. This tended to mean that the distributor was fully advanced by 2000 rpm. On today's gas, and in street applications, this is too fast. You would like to slow this advance down to where it is fully advanced at perhaps 3500 rpm. On some engines, this could be 3000 rpm and on others it could be 4000 rpm. There are aftermarket distributors available from Mr. Gasket, MSD, Mallory and Accel. Most of these manufacturers also offer springs or spring and weight kits for their distributor. MSD lists five different spring kits to adjust the curve from full advance at 2400 to 5500 rpm. Most distributors use two springs in the advance mechanism to determine the advance curve and these springs can be light, or heavy, two of each or a mixture. Some fast-advance distributors only use one spring.

This is a digital ignition box which still uses an electronic distributor. It is made by Crane.

This is another digital ignition box which can be used on street or strip. The digital ignitions are very high tech and offer many features.

The spark advance for the engine is built into the distributor. It is called the spark advance curve because it changes with speed or engine rpm. In the performance era, when good gas was readily available, you wanted to push the advance curve to come in very quickly—full advance by 2000 rpm—the upper curve. Now, you do not have good gas and on current pump gas you want to slow the advance curve down, especially at 2000 and 3000 rpm. The older production-type advance curve (lower curve) is closer but it could probably be fully advanced at 4000 rpm. This is done using more initial spark and a lighter first spring.

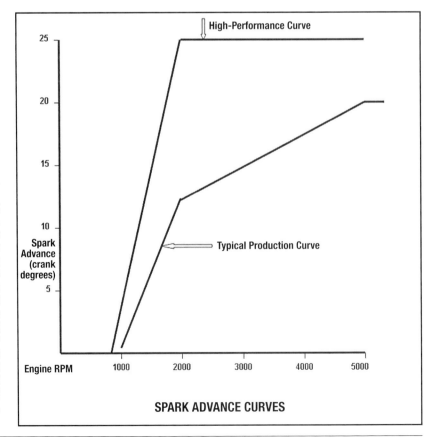

SPARK ADVANCE CURVES

Chapter 8
Carb Emissions, Fuel Economy & Power

There has been a lot of engine development over the years on engines using carburetors. Most of this development has been directed at improving horsepower and to a lesser extent, torque. Much of the basic horsepower hardware may also help fuel economy. Emissions is the new kid on the block and much of its performance development has been done on newer versions of the V8, but much of this technology can be applied to older muscle car era engines. Engine dynos are the source for much of this information and the key to the teamwork solution.

It may seem a little odd to include fuel economy and emissions with power but in many cases they can share much of the same hardware and benefit from the same technology. An emissions discussion is rarely included with any kind of performance consideration. Emissions regulations started in about 1967 with the PCV valve and the AFB and AVS were part of the team. The next major step occurred in 1972 but only the Thermo-Quad took part in this phase. A lot of the performance production engines and options ended at the end of 1971, including the AFB 8-barrel 426 Hemi, and many enthusiasts felt that the new emissions rules were responsible, which was only partially true. By the end of the 1970s, most of the big-block engines were gone and again emissions tended to be blamed. As production emissions standards tightened during the 1970s and '80s, most performance parts were sold as off-road, basically meaning that they were intended for some form of racing, not the public highway. By the early 1990s, there were a lot of pressures on the basic performance aftermarket manufacturers because there was a huge market of newer cars that they could not legally sell parts to or design performance parts for. In the mid-1990s, SEMA convinced the CARB—California Air Resources Board—to define a test procedure by which aftermarket performance parts could be approved and sold for street applications on current production vehicles. This test process, which gets the parts an emissions-exempt certification, opened the door to a flood of street performance development. Unfortunately carburetors had disappeared from production nearly ten years before this, but there is no reason that a performance carburetor package can't benefit from the testing procedure.

EMISSIONS

In the late 1990s, Mopar Performance developed a five-part performance package for their 360 Magnum V8 engines, basically in trucks. Each individual part was tested and received its exemption number from the CARB. All five parts were installed as a package and the complete package also passed the emission test. This complete performance emissions package gained something around 20–25% in power. The second stage of this program would have gained an additional 20+% in power, and while the cylinder heads, intake manifold and various hardware were introduced as performance parts, it was not tested for an emissions exemption. These two stages installed together would have offered emissions-exempt engines with about one horsepower per cubic-inch.

All the development on these emissions-exempt performance packages was done without the use of an actual carburetor. The main reason for this is that the emissions

REBUILD & POWERTUNE CARTER/EDELBROCK CARBURETORS

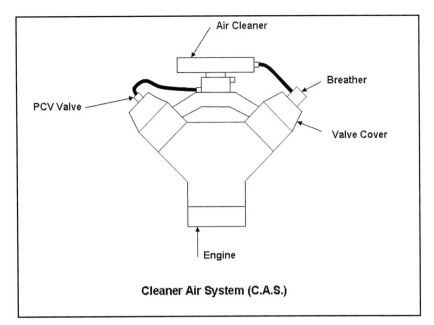

The first emissions package or Cleaner Air System (C.A.S.) is based on the PCV valve or positive crankcase ventilation. Most of the production engines built in the muscle car era used a PCV valve.

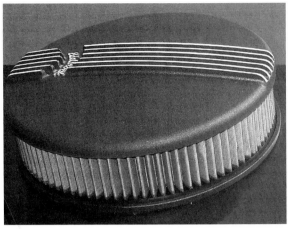

Many of the performance emission packages start with a low-restriction air cleaner—open face, 14" round, and high-flow, low-restriction element.

development started in the mid-to-late 1990s and all the production engines at that time were fuel injected. All of the original equipment manufacturers (OEMs) were involved in these emission development programs and the OEMs tended to take the lead. By the time that the testing process and procedures had been fine-tuned, no one wanted to go back and do carbureted engines built 15 to 20 years earlier. In most cases, it isn't a matter that the carburetor can't meet emissions but no one has tested a specific package based on a carburetor. There is at least one exception and that is the 50-state legal Edelbrock AFB carb (#1400) which is approved for all GM V8s built in 1980 and earlier.

This situation is similar to the problems that existed in the restoration market. The tooling for the parts was gone and it was expensive to build new parts that only fit old cars. After twenty-plus years, the demand builds up and old, used parts keep getting more expensive, so the aftermarket, in cooperation with the OEM, starts making individual replica parts for old cars. As they become successful, more parts are made and finally they make the whole car—sheet-metal and all. So far, GM and Ford are reproducing the 1968–'69 Camaro and the 1967–'69 Mustang complete bodies through the aftermarket, and more classic car bodies will be done in the future. To date, no manufacturer has elected to reproduce and recertify a '69 350 Chevy engine, carburetor and all, but it could happen.

Emissions Stages

With power, you have the basic horsepower scale to use for comparisons. With fuel economy you have the basic miles-per-gallon scale to use for comparisons. With emissions, you don't have a common scale to use, so in the three Stages that I have outlined in later sections, I have assumed a baseline emissions test and then upped the performance of each stage, holding the emissions at this same level. The key is to get all of the parts to work together, as a team. The performance gains for each stage could be as high as 20% but with the three stages, it is probably closer to 15% or 20% for the first stage and 10% for the second and third stages.

Ratings

One of the tricks to estimating performance is starting with a good baseline. To estimate your engine's performance based on a 20% gain means that you have to know accurately what the baseline number is. Most of the OEM ratings on 1972 and newer engines can be used as a good baseline. This isn't true on engines built in 1971 and earlier. The 340 four-barrel and some 350 Camaros are exceptions to this general rule. To try to help you estimate your engine's performance, try the guideline in the chart on page 153.

Let's assume that you have a 385 hp rated, 454 engine from 1969, the rated horsepower per cubic inch is 0.85 (385/454) so I would not use this for your estimates.

Note: For this example, use 0.75 to 0.80 factor from the chart on page 153. Now let's assume that you have a '69 350 that is rated at 265 hp at 10-to-1 compression ratio. This calculates to 0.75 and

Carb Emissions, Fuel Economy & Power

The Stage I package starts with a low-restriction air cleaner that assumes you do not have a hood scoop.

HORSEPOWER RATINGS & ESTIMATIONS

Engine Rating	Compression Ratio	HP/CID	Recommendation
Pre-1972:	all	0.85 to 0.90	Overrated, do not use
Pre-1972:	10-to-1 CR	0.75 to 0.80	OK to use
Post-1972:	9-to-1 CR	0.65 to 0.75	OK to use

The second part is a smooth, high-flow set of cast iron manifolds ('68–'70 383/440 versions shown).

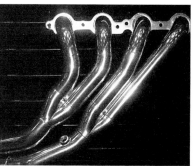

Optional: If you do not have access to a set of high-flow cast iron manifolds, then you can use a set of Tri-Y headers. The 4 tubes join into 2 pipes about halfway—center of photo—and then the 2 tubes join into 1—the collector.

The PCV valve (lower right) fits into the valve cover and the hose connects to the base of the carburetor.

The other part of the PCV system is the breather (arrow) that fits into the opposite side valve cover and its hose connects to the air cleaner.

should be OK to use. If you have a 9-to-1 350 from 1968 that is rated at 240 hp, it calculates to 0.68 and should be OK to use. There are many potholes in this estimation equation. For example, if your production engine has some very bad (performance-wise) exhaust manifolds, then just changing the exhaust manifolds may result in giant gains.

The point of this is that these emissions-exempt engine packages were developed for street use and there is a lot to be learned from this really high-tech development that can be used for your high performance street package. Perhaps the real key is to make a package and all of the parts have to work together so you should look at it as a team.

For muscle car engines built in the '60s or '70s there is very little to use as a baseline since there are so many options and possibilities. So I have selected some typical parts but they have not been tested and are listed only to give you a package that might have a good chance at passing an emissions test (with improved performance).

Note: The parts have been tested individually but not put together as a package for this application.

Stage 1

Parts—PCV and breather, high-performance air cleaner, cat-back exhaust, Tri-Y headers, high-performance cam and springs, 50-state legal carb, electronic ignition.

Why?—The system starts with the PCV and breather, which means that the valve covers must have grommets to allow these parts to be installed and function properly. You must use an electronic ignition because all the upgrades and high-performance designs are based on some version of the electronic ignition, plus a point-based system requires almost constant adjustment. The Edelbrock 50-state legal carb gives a good 600 cfm starting point for four-barrel package. A high-performance air cleaner is a low-restriction element that is usually 12"–14" in diameter with open face. The cat-back exhaust system is a low-restriction muffler and large diameter tailpipes, even if the car does not have a catalytic converter. Tri-Y headers are an option because you could use high-performance cast iron

Rebuild & Powertune Carter/Edelbrock Carburetors

Instead of adding a hood scoop, the cold-air system is done by using a smooth, molded plastic tube (to the right) to connect the low-restriction air filter to the top of the carburetor. This air cleaner is mounted remotely—typically down low next to the radiator so it picks up the cold air coming under the car and under the radiator. This may require some research and some adapting because they tend to be made for newer engines, not muscle car era models.

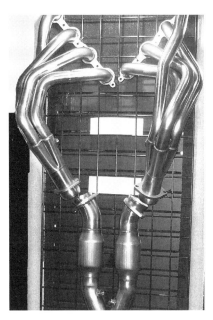

Replace the cast-iron manifolds with 4-into-1 headers as shown. A Tri-Y design is your first choice, but they are hard to find unless you have them custom made. Then add an aftermarket catalytic converter, bottom of photo, one per side.

You should have a good, high-rise aluminum dual-plane intake manifold for a four-barrel carburetor. Some high-performance production manifolds (cast iron) are OK.

When carburetors were being used on production engines, there were very few options for cylinder heads. Today, most of the popular V8 engines have aluminum and cast iron cylinder head options available. Try to keep the intake port volumes down, close to production and valve sizes close or equal to the high performance production versions.

exhaust manifolds. Most standard production engines use cams with 0.400" lift, so you want to move up to a high performance cam, which means around 0.450" lift, but keep the centerlines wide—114–116 degrees. It is always a good idea to replace the valve springs on older engines even if they are heavy-duty springs.

Stage 2

Parts—Bigger AVS carb, bigger cam, high-performance dual-plane aluminum intake, cold-air system, high-performance coil, catalytic converter.

Why?—Moving up to the next level in performance means you have to tune the emissions side of things. Start by adding an aftermarket catalytic converter. If you have a converter already then it is probably okay. The original Stage 1 carb is fine for small-blocks but I'd recommend moving up to a 750 cfm unit for big blocks. Assuming that the Stage 1 package used a production cast iron manifold, I'd recommend moving up to an aluminum high-rise dual-plane intake manifold. A cold-air system is stolen from the newer cars/trucks that use a molded plastic part that attaches to the carburetor and picks up cold air low behind the radiator.

Note: Cold-air systems are made by manufacturers like K & N, Airaid Filter Co., and AEM (Advanced Engine Management). A high-performance coil is added to the base electronic ignition like an MSD Blaster unit. The bigger cam should be in the area of 0.470" to 0.480" lift.

Stage 3

Parts—Hydraulic roller cam, high-flow heads, max compression ratio, charcoal canister, re-curved distributor.

Carb Emissions, Fuel Economy & Power

If you have a sealed bowl vent that vents to a fitting rather than the atmosphere, then you could consider adding a charcoal canister, like this one from Year One. I would not recommend adding this to a production AFB or AVS since they vent to the atmosphere.

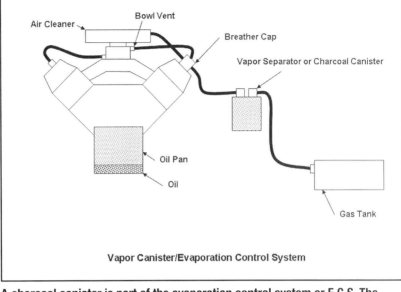

A charcoal canister is part of the evaporation control system or E.C.S. The canister is connected to the breather in one direction and the gas tank in the other.

Setting the timing properly is as important to emissions as it is to performance. Use a high-performance–based electronic ignition and a high-performance coil. See Chapter 7, page 150, for a tip on recurving the distributor.

Centerlining the cam is very important. It requires a degree wheel, lower part of photo, and a dial indicator, top. If the cylinder head is on, you use a TDC positive stop to locate the piston accurately, in case the damper has slipped.

Why?—The big change here is the high-flow performance heads, but there are aluminum and cast iron versions that fit here. Keep the valve size down to high-performance specs and keep the ports small or low in cross-sectional area. Many engines built in the 1970s and '80's have only 8-to-1 compression ratio so you can push this to 9-to-1 on pump premium gas. Adding a charcoal canister can be tricky if not installed in production. Now you would like a bigger cam to go with the hi-flow heads and you can use a 0.500" lift hydraulic cam with wide centerlines but you could also install a hydraulic roller cam with similar lifts.

General Upgrade

High-performance spark plugs, 7.5mm plug wires, synthetic oil, low viscosity, slow water pump and alternator, windage tray and viscous fan,

Why?—There are many items in the engines that relate to performance that can be helpful to any package. A viscous fan and a windage tray increase the engine's output at every engine speed, making the engine more efficient. As long as you do not go too far, slowing the water pump and alternator down by using a smaller crank pulley made of billet aluminum typically allows the engine to make more power at all engine speeds. High-performance spark plugs can offer many advantages, especially with performance ignitions. Older engines need new spark plug wires with a larger wire diameter (7.5mm) and more insulation, which helps the high-energy ignition do its job. I would also recommend synthetic oil like Mobil One, because it can be easily purchased with lower viscosity like

EMISSIONS-LEGAL EXHAUST SYSTEM

A legal exhaust system really means an emissions-exempt exhaust system (related to CARB—California Air Resources Board), which is overall generally a stricter test standard than federal standards. This means that the manufacturer had his system tested on the proper vehicle and the part did not increase the engine's emissions. So how does this affect the performance four-barrel engines with the carburetor, since they were last built in the 1980s during the second emission phase?

Technically speaking, the '60–'71 muscle car four-barrel engines only have to meet 1971 emission standards, which means cast iron manifolds, muffler, PCV and breather, etc. Let's assume that you wanted to install a high-performance muscle car four-barrel engine into a newer car and try to meet the newer exhaust emission standards, which relates to green concerns. To do this, you would want to try to copy the late-1990s high-performance parts packages designed for cars and trucks that received emission exemptions, but the key would be the catalytic converter similar to those used on newer cars. Today, the aftermarket offers high-performance catalytic converters with large inlet openings, like 2 1/2" (recommended for engines over 400 cubic inches), but the small-blocks (360 and smaller) could use 2 1/4". Larger displacements, perhaps 500 cubic inches and up, might want 3" inlet openings. Then you want to use new car/truck headers, and typically these headers are a Tri-Y design that offers maximum engine torque with the best power available.

Note: Most shelf-headers for muscle car engines are 4-into-1 headers, so a Tri-Y design would have to be custom made or adapted.

Note: Tri-Y headers returned to popularity in the mid-1990s with the newer production vehicles, especially trucks, because the Tri-Y design offers a wider powerband and more torque. Then you add the cat-back exhaust system, which replaces the muffler and tailpipes.

Note: Cat-back terminology comes from newer cars that had a catalytic converter, in many cases more than one, and a replaced exhaust system that was rearward of the production catalytic converter to the rear of the vehicle. You want to use a 2 1/2" to 3" cat-back system. If possible you also want to use an H-pipe, located under or slightly rearward of the transmission crossmember.

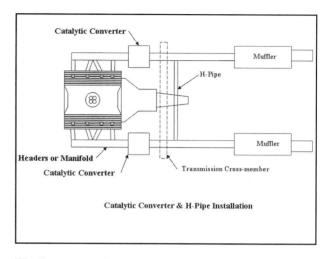

Whether you use headers or cast iron manifolds, you should add an H-pipe behind the collector or catalytic converter. This pipe is usually added near the transmission crossmember.

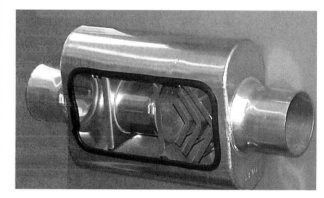

Low-restriction mufflers, compact in size, are much more readily available today than they were in the muscle car era. Large diameter inlets and outlets are also available. This is a Flowmaster design.

Power Gain

To actually estimate the power of an engine, it should be run on an engine dynamometer. Most of the parts discussed earlier were first run on newer engines that had used a fuel-injection system. Since the general airflow capabilities were about the same, the engines should respond in similar manners. New technology on older engines is somewhat like free dyno time.

FUEL ECONOMY ADJUSTMENTS

In the muscle car era, fuel economy was not a consideration. The price of gas was cheap and everyone wanted more power. In the 1980s and '90s with gas prices in the $1 and $2 range, it was still not much of a concern. Then in the mid-2000s gas prices worked their way up to $4 per gallon and customers noticed. Today fuel economy is a big

5W-30. This lower viscosity will help cut engine friction at all speeds. Performance oils are available from suppliers such as ExxonMobil and Lucas Oil Products.

With no tests or dyno development available, it is only a place to start. With so much interest in green topics these days that directly relates to emissions and fuel economy, this shows that performance and carburetion can be in the picture too.

Carb Emissions, Fuel Economy & Power

You may find it hard to believe that the cooling system can affect power and economy, but it does. Drag racers use cooling systems that work for very limited distances but this is not recommended for street or dual-purpose engines. Typical production cars overdrive the water pump and alternator by 20% to 25% by having the crank pulley larger than the water pump pulley. Performance aftermarket billet aluminum pulley makers like March Performance Pulleys reverse this and slow the water pump and alternator down by making the crank pulley (lower one) smaller than the water pump pulley (the upper one). Caution: If you do this and additionally use a very low idle rpm, the alternator may discharge at idle.

For performance, colder thermostats have always been recommended. For fuel economy, you should use a 195°F thermostat rather than the 160°F or 180°F units.

If you have a lot of accessories to drive, you might consider an aftermarket kit to install a serpentine-belt system as shown.

consideration. If it is a big deal for new car customers, then it should be of interest to the aftermarket customers as well. However, I think that the general feeling of the performance aftermarket enthusiast is that you can't have both, which isn't completely true. Muscle cars had 400 cubic inches or more, eight-barrel carburetor setups and big, high-overlap cams and 4.10 gears. They went fast but no one would consider them a fuel economy package because the customer had no interest. The emissions development done in the 1970s, '80s and '90s taught us that you can have both—emissions, fuel economy and performance. Much of the new aftermarket hardware that helps performance can also be used to help fuel economy.

Carburetor Adjustments

Obviously you can run the A/F ratio one step lean but it is not a good idea to push too far in the lean direction, because you take the chance of detonation and higher temperatures and hurting the engine. Remember that you can go slightly leaner on the primary and then setup on the secondary jet to keep the same total fuel at WOT.

Inlet System

A dual-plane intake manifold is the best choice for fuel economy. If you do not have a hood scoop, you would like to have one of the new-style cold-air inlets that are made of molded plastic and are designed to pick up the air below the radiator on one side.

Cooling System

Using a small crank pulley is good for fuel economy because it slows down the water pump and alternator.

Caution: If you plan on cruising, do not go too far or the engine may not charge the battery at low engine speeds or it could overheat. Billet aluminum pulleys are available from March Performance Pulleys. You probably want to select a size about halfway to the drag race package. Another tip is to use a viscous drive on your fan. Additionally, the drag race package usually used a 180°F thermostat in the cooling system or no thermostat at all (race only) and for fuel economy, you would like to use a 195°F thermostat. In many applications big radiators can help the cooling. These are available

Bigger oil pans offer free performance but are always a compromise for any street vehicle. The general street conditions limit the depth, so you then make the pan's sump, the bottom part of the pan where the oil collects, wider that the basic pan itself.

from Be Cool Radiators and U.S.Radiator, among others.

Oiling System

The easy one in the oiling system is to use a windage tray below the crank. A good oil pan is also a plus, just like it is on the power side. I would recommend a synthetic oil but one with less viscosity than usual—perhaps a 5W-30.

Windage Losses—When the crank, rods and pistons spin around in the crankcase, they have to drag themselves through the oil that is sprayed everywhere inside. This is measured as *windage losses*. The windage tray cuts the losses and therefore results in an output gain. If you can eliminate a loss, it results in a gain. You might consider this free horsepower. Less viscosity in the oil also fits into this category. Another windage gain, especially on small-block engines, can be found by using bigger oil pans. The trick is to add a larger oil pan without compromising ground clearance. For street vehicles, the extra size should be in the width rather than the height. Larger oil pans are available from suppliers like Charlie's Oil Pans, Canton Racing Products, Moroso and Milodon.

Ignition System

There are some high-tech spark plugs that can be used, such as the Bosch 4-electrode platinum units. You should upgrade the spark plug wires (perhaps 7.5mm) to run more energy from the ignition/coil/plug system. You definitely must have an electronic ignition system. Use a vacuum-advance distributor and max-power spark advance and a higher output coil, like a Blaster from MSD. Remember the solution for the high-load, low-rpm problem that was discussed in Chapter 7 on page 150. That may have to be done for fuel economy as well.

Coatings

There are many coatings that are available for the various parts of your engine. However, all coatings are not the same. Some coatings block heat and

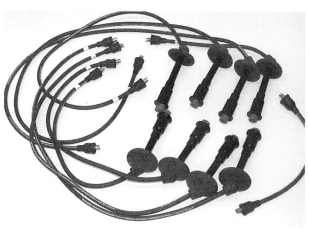

As you increase the ignition system's energy and output, you need to install better spark plug wires. I recommend 7.5mm premade units and you would like to have each one cylinder numbered. Additionally, try to keep the coil wire, far right, as short as possible.

some are designed to lower friction. Others are designed for use on headers. There are many suppliers for these coatings like Swain Tech Coatings, Polydyn Performance Coatings, HPC Performance Coatings, Jet Hot Coatings and Calico Coatings.

Exhaust System

Less backpressure is desirable, so you want one of the latest cat-back exhaust systems adapted for your car. That means a high-tech, low-restriction muffler (dual exhaust if possible and an H-pipe) and larger diameter tailpipes. If available for your car, get larger exhaust pipes that connect the exhaust manifolds to the mufflers. Upgrade to a set of high-tech, Tri-Y headers, or a set of high-flow, high performance cast iron exhaust manifolds will work very well and tend to be quieter than tubing headers. Consider suppliers like Borla Performance, Flow Master, tti (Tube Technologies Inc), Hedman Headers, Dynatech and Burns Stainless.

Cams and Valvetrain

Production street cams are very mild. The aftermarket makes some high-energy designs in the 0.450" to 0.500" lift area that can make a lot more power and still give decent fuel economy but the trick is to use a wide cam centerline which is ground into the cam by the manufacturer.

Note: This style of cam is not always an off-the-shelf item.

The specifics of a cam are very important to all aspects of engine performance but even mechanical and hydraulic cams look the same in a picture. Try to use street cams that have wide cam centerlines like 114–116 degrees.

In the muscle car era, displacement was limited by production hardware, but today long stroke cranks like this forged unit offer large increases in displacement. Special high-performance cranks are available from Scat Industries. Today, even larger-bore capacity blocks are offered, even in aluminum, from suppliers like Keith Black Racing Engines.

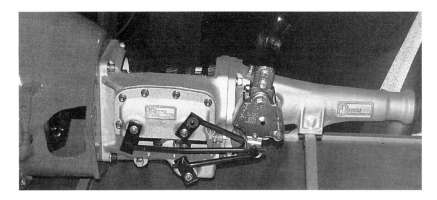

Automatic transmissions are by far the most popular choice in drag racing. However, manual transmissions will get better fuel economy. A few years back, many people had to remove manual packages because they couldn't get parts, but that is not the case today. New aluminum case parts are available for the A833 4-speed transmission (shown).

Cylinder Heads

Cylinder heads can be expensive but there are so many options today that you will have a hard time making a selection. The key is not peak flow but velocity. You want to keep the velocity up in the port but this information is not published. Intake ports with large cross-sectional areas will have low velocity. You would like to stick with the stock performance valve sizes and keep the port volume down. Assuming that you are going to use a 0.500" lift cam, do not look at intake port flows above 0.500" lift. Ports that are designed to flow at 0.700" and 0.800" lift are probably too big and come with big valves. With limited information, another trick to find what you need is to look at the flow at half lift or 0.250" (based on my 0.500" example). Therefore select the head that flows the best at 0.200" or 0.300" lift. Most manufacturers will publish the full lift curve but they only list the 0.2 and 0.3 type numbers in the typical flow chart. There are many, many suppliers of cylinder heads including the GM, Ford and Mopar performance parts divisions and Brodix, RHS (Racing Head Service), Airflow Research, All-Pro Aluminum Heads, World Products, Indy Cylinder Heads, and Stage V Engineering.

Cam and Head Package—When you want to flow air at 0.700" and 0.800" valve lifts, you make the ports larger and use large valve head diameters. Larger ports (in cross-sectional area) and large valve head diameters will negatively affect the port's velocity. If you install a bigger cam to work with 0.700" and 0.800" ports, the durations must increase from what you can use at 0.500" lift and these longer durations impact low-speed torque and general throttle response, so you now have to use a higher stall converter and more rear gear. At this point, you now have basically a race car.

Displacement

Obviously, more cubic inches tend to use more fuel. Today there are so many easy displacement packages that engines with extra displacement are readily available. One approach for using more displacement is to combine it with an overdrive transmission. Properly built, the large displacement will make more torque so it can run less engine speed (rpms) at highway speeds and still offer good drivability. The lower engine speed can help balance the extra displacement.

Manual Transmission

There are four-, five- and even six-speed manual transmissions that can be used. The top gear in these units can be an overdrive unit which can generate better fuel economy. Using the extra gear in the transmission as the overdrive and combining that with a 3.91 or 4.10 axle ratio can give you more acceleration and more fuel economy in the same package. Axle ratios are available from Auburn Gear and Richmond Gear.

Note: There are overdrive transmissions and add-on overdrives. The OD ratio tends to be 0.69 or 0.70.

FORMULA FOR MPH

The basic formula for calculating your vehicle's mph based on your tach's engine rpm reading is as follows:

mph = (td x rpm)/(336 x gr)

mph = vehicle miles per hour
td = tire diameter measured in inches
rpm = engine rpm tach reading
gr = rear axle gear ratio
336 = a constant but does not account for tire growth or converter slip in automatic transmission cases.

This basic formula can be rearranged so you can calculate your engine's rpm for various changes, which may be more helpful for fuel economy considerations.

rpm = (gr x mph x 336) ÷ td

Let's look at an example—what rpm is the engine turning at 60 mph using a 28" diameter tire and a 4.10 rear axle ratio? rpm = 4.10 x 60 x 336 ÷ 28 = 2952. If you add an overdrive transmission into the equation, then the final or overall gear ratio drops. The typical overdrive ratio is about 0.70, so the final drive gear ratio would be 4.10 x 0.7 = 2.87. The new rpm at 60 mph is 2066 rpm. That's almost a full 1000 rpm off your cruising rpm, which makes much better fuel economy. Sticking with this package, if you change the tires to ones that have a 24" diameter, the 60 mph rpm goes up to 2410 rpm. On the other hand, if you install tires with a larger diameter like 30", then the rpm drops to 1928 rpm.

The key instrument in the mph formula is your tachometer.

The key piece of hardware in the formula is the rear axle ratio.

There are also kits to install overdrive automatic transmissions into various vehicles like the Keisler unit.

There are many suppliers for torque converters. This TCI unit is cut apart for display. Make sure you get the right torque converter for your application. What works for a race car does not work for a street or dual-purpose car.

Note: A833 parts are available from Passon Performance and overdrives are available from Keisler Overdrive and Gear Vendors Overdrive.

Automatic Transmission

Older muscle car automatic transmissions were usually two or three speeds. Today there are four-speed ones available. There are also adapter kits that are available for many engine choices. An automatic package could also go with one of the lock-up style automatics, but kits aren't as readily available. To use the typical lock-up automatic transmission, you would have to change both the transmission and the torque converter. Another choice is a tighter (lower stall) torque converter, but efficiency information on aftermarket converters is very limited. More efficient converters tend to be larger in diameter than race converters—11" or 10" rather than 8" or 9". Special torque converters are available from TCI, ATI, Turbo Action, Coan Engineering, Transmission Specialties, and Pro/Race Performance Products.

CARB EMISSIONS, FUEL ECONOMY & POWER

The aftermarket is making parts for many different engines like this 348-409-427 W-Series Chevy.

While many crate engines are shipped without a carburetor, the ready availability of Edelbrock four-barrel carburetors makes them a very popular selection.

Horsepower comes in many shapes and sizes like this 426–472 Hemi crate motor.

In you are planning a large displacement project like a 408-inch small-block Mopar, a crate motor can be a good selection because you get all the parts together and its all assembled and ready to run.

Note: There are overdrive automatic transmissions available and the overdrive ratio is 0.69 or 0.70. Overdrive kits are available from SMR Transmission, B & M and Keisler.

POWER

Actual horsepower and torque numbers are determined through dyno testing. This can be expensive. You can also wear out your brand-new engine. You want parts that work well together and one way to find this information is to copy the highly developed OEM crate engines. With any of the factory crate engines, engineers have spent a lot of time on the dyno finding out what works best together and built it into an assembly, which you can purchase or copy since the factory tends to publish all the specifications and data, perhaps even publish a horsepower/torque curve, rather than just peak numbers.

There are many books and articles on each specific engine and how to make horsepower. The common denominator in all horsepower tests is BSFC which stands for brake specific fuel consumption. Almost all naturally aspirated engines on normal gasoline have a BSFC of about 0.5 lbs/hp. This basically means that the more horsepower that you make, the more fuel that you use. One approach to helping fuel economy while trying not to hurt your wide-open throttle performance is to work on the low rpm and part throttle side of the curve. This can require some dyno time and could be expensive.

Note: The OEM crate motor is not always the same size as an actual production engine. For example there are 528 cubic-inch Hemi crate engines and 502 cubic-inch Chevys and 408 cubic-

Most rear-wheel drive vehicles have somewhat of a nose-heavy weight distribution. Drag racing wants to put as much weight on the rear wheels as possible. For general street or dual purpose applications, you would like to have a 50-50 balance. For the typical rear-wheel drive V8 car, this means moving some weight to the rear like putting the battery in the trunk as shown here.

inch Mopar small-blocks just to name a few. Obviously you can also select to have your engine built by a professional shop like Ray Barton's Racing Engines or Koffel's Place.

One of the aspects of dyno testing that you will notice with high performance parts is that they will make more power but many tend to make this power peak at higher rpm. Using more rpm means using more fuel, so this isn't the direction that you want to go. What you would like to do is make more torque and keep the engine speed down, perhaps not too far from stock-type rpm. This way the high-torque engine will pull a lower (smaller number) rear axle ratio and generally lower stall converters and overdrives if they are installed. The two aspects working together can make a team that goes faster and makes better fuel economy.

CHASSIS

Most enthusiasts will work very hard on the engine but don't usually put as much effort into the chassis. Many of the items that you work on in the chassis to make the car faster also help with fuel economy. Less total vehicle weight helps both aspects. Weight distribution or balance is defined as the amount of the total vehicle's weight that is on the rear tires. For performance (racing) you want as much rear weight distribution as possible while a maximum of 50% is probably ideal for economy because it equalizes the rolling resistance of the four tires.

Note: It is hard to get 50/50 weight distribution on a typical rear-wheel drive V8 production-based car. The rolling resistance of the tires is a big factor in fuel economy but not as much in acceleration contests. Rear slicks are great for drag racing but have a very high rolling resistance. Something as simple as moving the battery to the trunk helps improve weight distribution both front to rear and left to right.

Drag racers spend time on their front end alignment to be sure that the wheels are pointed straight ahead when the vehicle goes through the traps and this is just as important on street cars/trucks or for power or fuel economy. If you have an older car, muscle car or even a 1970s or '80s rear-wheel drive car, most of the bushings in the front end will be worn so that when you have the front end aligned, everything is fine, but after a few hundred miles of street driving, the bushings allow movement and nothing is aligned anymore. Replace the bushings in older cars when you do an alignment. New bushings hold alignment much longer.

Chapter 9
Racing & Special Applications

The 1966–'71 426 Street Hemi was the most popular Hemi and longest-lasting dual four-barrel production inlet system but the Hemi cross-ram is more popular in racing today.

The basic carburetor has been used in racing as long as there has been competition. In the muscle car era, carburetors were raced on all doorslammers drag racers, including Stock and Super Stock cars. Even though they have not been used in production for almost 20 years, carburetors are still the most popular induction system used in racing. By far the most popular carburetor is the basic four-barrel unit. Today, these readily available four-barrel carburetors come in 500 cfm to 800 cfm sizes. You can use two of them and double your flow to 1000 cfm and 1600 cfm respectively. Refer to Chapter 3.

RACING FOUR-BARREL CARBS

There are many four-barrel carburetor options available today. Although the two-barrel carburetor may have been the most popular the last 20 to 30 years of OEM production, it wasn't very popular in racing. Most were too small to deal with increased horsepower output and higher rpm. There were some 350 and 500 cfm two-barrel carburetors available, but these have limited use. The typical performance engine is a V8 with between 300 to 500 cubic inches with larger versions more likely than smaller ones today. With engines of this size the readily available 500 cfm to 800 cfm four-barrel carburetors are just about perfect.

Manifolds

Most four-barrel intake manifolds can be found in both dual-plane and single-plane versions. The height of the manifold can affect hood clearance because you do not always want to be forced into using a hood scoop. The high-rise manifold style generally makes more power than low-rise versions.

Obviously for racing you can use two carburetors to double your airflow. Race engine builders prefer one carburetor—usually a single four-barrel—rather than two or

The most popular racing intake manifold is the standard four-barrel, single-plane.

three carburetors. Additionally, many racers prefer to dial the car back (slow it down) and use an air cleaner. The exception is Stock and Super Stock engines, which do not use an air cleaner—the most ideal setup (max flow).

Using a four-barrel carburetor in racing generally means that the engine will be making more power and probably will be using more engine speed, rpm. This means that the jets and metering rods in the carburetor will have to flow more fuel into the intake manifold but with the wide selection of tune-up/calibration parts offered by Edelbrock, richer fuel mixtures are not a problem. A drag race engine only has to be concerned with idle and wide-open throttle applications, so you can follow the calibration steps outlined in Chapter 7 and adjustments from Chapter 6, as well as the jet size calculation in Chapter 4.

While still a single plane, this manifold would be considered a race single plane because the runners extend into the plenum. These manifolds are also taller and use runners with larger cross-sectional areas. This one is for the small-block Chevy.

This race single plane is designed for the Vortec Chevy. Note the vertical intake manifold attaching screw holes.

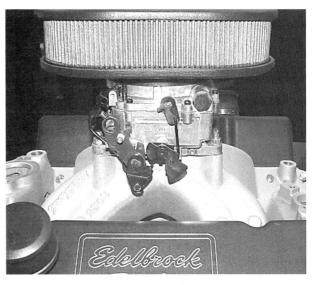
This is also a race single-plane intake manifold, but it is mounted on an engine. You can see how high above the valve cover that the carburetor pad sits. This taller manifold can cause problems with hood clearance or require a hood scoop..

This is an air gap manifold for the small-block Ford. In this design, air passes under the plenum and runners from front to rear and can also enter/exit between the runners on the side. The bottom plate keeps the hot oil off the bottoms of the runners, delivering a denser charge to the cylinders. Air gap designs can be either dual-plane or single-plane, but the dual-plane versions are harder to make.

This is a race single-plane four-barrel intake manifold, and you can easily see how much taller it is. A standard manifold might only come up to the bottom of the plenum of this design.

High-Rise Racing Four-Barrel

NHRA/IHRA Replacement—Perhaps the first racing carburetor concern is a replacement or superseded number. The NHRA/IHRA Stock and Super Stock classes must have a legal carburetor. In some cases, the new Edelbrock carburetors have been superseded on specific engines, and in some cases they have been accepted as a legal replacement. This aspect of racing is constantly changing, so I would recommend contacting the NHRA/IHRA Technical Department directly. The '66–'71 426 Hemi Street Hemi engine is allowed to use the Edelbrock carburetors as a replacement.

Manifolds—There are many options for four-barrel intake manifolds. Street rods, street machines, bracket race cars and dual-purpose vehicles each have unique requirements that the manifold and carburetor have to meet. The intake manifold also has a direct impact on the height of the assembly over the valve covers, which relates to hood clearance and hood scoop requirements.

Advantages—You need to consider similar reasons for using a carburetor in racing as you did your basic selection—airflow that ranges from 500 to 800 cfm, costs, general parts availability especially relating to mixture adjustment capability, adjustability and is the carburetor user-friendly. One added aspect for a resto-based project is general appearance and does it look like the original.

Eight-Barrel Racing Carbs

The basic eight-barrel carburetion setup consists of two four-barrels working together. It was quite popular in the late 1950s and early '60s. Engine displacements had gone above 400 cid, and the stock single four-barrel carburetors flowed around 400 cfm, so adding a second carburetor was an easy way to get twice as much airflow to feed the larger engines. By the mid-1960s, all of these production-based eight-barrel system were gone except for the 426 Street Hemi which made it through 1971. The impressive appearance of multiple carburetion especially two four-barrels was also a major selling point of increased performance. This aspect remains in the judging of various engine packages in the street rod and street machine car shows.

RACING & SPECIAL APPLICATIONS

This is the stock dual, four-barrel inline system, carbs and intake manifold, used on the 1966-'71 426 Street Hemi engine. The intake manifold is a dual-plane, and has been re-cast by Mopar so it is available new.

This is an eight-barrel inline system with polished carbs and intake designed for the small-block Chevy LS1.

An eight-barrel inline carbureted intake system, carbs and manifold, for the 5.7L and 6.1L Mopar Hemi engines. Note: All production versions of this engine in cars and trucks are MPI (multi-point fuel injected). However they are becoming very popular in street rods and street machines and carburetors offer many advantages in these applications.

Production—In the late 1950s and early 1960s, there were many production eight-barrel intakes. Most were made of cast iron and featured a basic single plane design. With two carburetors and eight throttle bores, the basic single plane design is the easiest eight-barrel manifold to make for use with wedge-style cylinder heads. The dual-plane design offers performance characteristics that better match the eight-barrel carburetion when used for street/dual purposes. Newer designed manifolds are better choices today but NHRA/IHRA Stock classes require the stock manifold while Super Stock classes can use any manifold of the same configuration which means most Super Stock eight-barrel manifolds are fabricated.

Chevy—In the 1950s and early 1960s, Chevy offered an eight-barrel system for their small-block engine and additionally offered one for the 348 and 409 engines, which was probably the more famous one at the time. Today, Edelbrock makes a new aluminum inline eight-barrel manifold for the small-block and also a new one for the 348-409 (the W-series heads).

Pontiac and Buick—Both Buick and Pontiac offered inline eight-barrel carburetion systems in the late 1950s.

Ford—Several Ford engines in the late 1950s and 1960s offered inline eight-barrel carburetion systems including the 289 V8 like the Shelby-American cars. Edelbrock makes an new aluminum inline eight-barrel system for the Ford small-blocks and for the Ford 390/406/410/427/428 cid big-block engines.

Mopar Wedge—There were two inline eight-barrel systems used on the Mopar wedge engines— one for the 361–383 big blocks and one for the 413-426 Raised-big-blocks. These manifolds were made of cast iron and were single plane designs. Edelbrock makes an aluminum, inline eight-barrel dual-plane manifold for the 413-426W-440 engines.

Mopar Hemi—The 426 Street Hemi was the last inline eight-barrel system used in production (1966–'71). It was an aluminum dual-plane design. It has been remade for the restorations so it is still available. There is also an eight-barrel carbureted intake manifold for the new Gen III Hemi (5.7L and 6.1L).

Race Inline Manifolds—Race inline manifolds such as those made by Indy Heads are designed for racing-only applications and feature very large intake runners and large manifold plenums. Typically they feature a two-piece construction which means the the top comes off and its held together with lots of small screws.

Cross-Rams—The cross-ram eight-barrel carburetion system is perhaps even more impressive than the inline arrangement. There have been several race-based cross-ram system such as the '69 390 AMX Super Stock package (50 car program) and some small-block Chevy packages but most of the production versions were done by Mopar. This started with the long-ram systems done in the late 1950s and very early 1960s, which had the carburetors actually sitting outside of the valve covers (wedge heads). These were the first tuned intake manifolds but they were designed for the 1960 engine package and are way too long for today's engine package, and tune for a relatively low engine speed, especially with today's engine hardware.

This is a eight-barrel cross-ram designed for the big-block Mopar engines. The carburetors are opened at the same time with the bell crank that sits between them.

This is another style of eight-barrel cross-ram. Note the plugs (2) visible in front of the closest carburetor and the one plug in front of the right carburetor. These plugs allow access to the intake manifold attaching screws. There are also 2+ attaching screws accessed through the carburetor openings. Access holes/plug like this are common on cross-ram manifolds because they are so large and take up so much space around the heads.

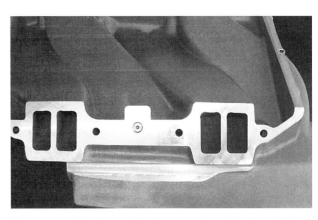

A cross-ram intake upside down to show the size of the runners. The production 413–426 Max Wedge cross-rams had very large tall intake runners and don't seal to the much more common 383–440 size ports. This A & A Automotive and Transmission version can be cast with either size ports—note the extra material around these ports at the manifold face.

The next step in cross-rams was the one-piece casting used for the 1962–'64 413–426 Max Wedge engines. It was used in production and was probably the highest volume cross-ram unit. The intake runners in the manifold were very large to match up to the big ports used in the Max Wedge heads and they didn't line up with the standard 413/426/W440 style heads (much taller port especially at the top). Today A & A Automotive and Transmission makes a version of their aluminum Max Wedge cross-ram with the smaller runner that will match-up to the standard 440-style ports. A & A Automotive and Transmission also makes a Max Wedge big-port version for the restoration/race applications.

The most famous of the cross-ram packages was used on the 1964, 1965 and 1968 426 Race Hemi engines. It is still being raced competitively in Super Stock today, which has become the most popular NHRA Super Stock class. The original intake manifold was a one-piece aluminum casting and the newer version was made of lighter magnesium. Mopar currently makes a 3-piece aluminum version (plenum tops are removable) and A & A Transmission makes a one-piece version for resto applications.

In the late 1960s and early 1970s, Edelbrock made several 2-piece, aluminum cross-ram intake manifolds. The Mopar 383 and 440 versions were called the STR-14 and 15. There was also a Mopar small-block version called the STR-12. The '69 AMX used another version. There were several versions made for the small-block Chevy. Edelbrock no longer makes any of these intakes (at this time).

Tunnel Rams—Tunnel ram intakes generally use two four-barrel carburetors mounted inline but the manifolds are very tall. This means that they stick through the hood and require a giant hood scoop. The first tunnel ram was used on the High and Mighty drag race car in 1959 using WCFB/AFB carburetors. With the creation of the new Pro Stock drag racing class in 1970, tunnel rams reached their peak in popularity. Almost all intake manifold manufacturers made tunnel rams in the early 1970s but few are available new today. Tunnel rams are a race-only system. Most of the tunnel rams being raced today in classes like Pro Stock are fabricated.

Fabricated—Today there are numerous sheet metal intake manifold fabricators.

Note: The sheet metal might indicate a steel material, but it is actually sheet aluminum. The construction technique is similar for both metals, however. A fabricated manifold is basically welded together from sheet aluminum instead of being cast of cast aluminum. The process is very good for very low-volume manifolds, or prototypes, but they are expensive. You can fabricate almost anything to test any idea and don't have any money tied up in tooling if it doesn't work or if further changes are indicated. Today, fabricated intake manifolds are legal in NHRA/IHRA Super Stock classes. Almost all cross-ram and inline eight-barrel manifolds used in SS classes today are fabricated but the majority of the single four-barrel manifolds are still cast.

Note: Any fabricated intake manifold requires dyno time to sort out. Fabricated manifold sources include Hogan's Racing Manifolds and Wilson Manifolds.

RACING & SPECIAL APPLICATIONS

The Edelbrock STR family of manifolds were 2-piece designs. This is the bottom of an STR manifold for the big block Mopar engines. Note that the manifold had to be installed with the lid off to allow access to some of the intake manifold attaching screws.

Crate motors come in many sizes and shapes. This one is a 426–472 Hemi from Mopar. The GM, Ford and Mopar OEM crate motors seem to be the most popular.

Edelbrock offers complete crate engines also like this eight-barrel small-block Chevy. Edelbrock versions also feature new parts, not rebuilt parts.

CRATE MOTORS

All of the major OEM factories (GM, Ford and Chrysler/Mopar) offer many versions of their crate engine assemblies. The major aftermarket manufacturers like Edelbrock along with some engine builders also offer crate engines. The OEM factories spend big bucks on the dyno development time sorting out the parts that work best with their engine and then sell it as an assembly. The biggest advantage is that these companies publish the data. Over the last 15 to 20 years, crate engines have become very popular. One part number gets you all the parts that you need and, for the most part, the parts are brand-new.

Some aftermarket manufacturers also offer their own crate motors like Edelbrock. While some engine builders may still be rebuilding blocks, Edelbrock is using new blocks available from GM/Chevy.

The nice thing about these factory crate engines is that they come with most of the specifications for the engine, including a parts list and even power curves in some cases. This way you can copy a well-developed, specific package for your application based on reliable data. Some companies have offered short blocks (no heads or intake manifold) but these packages haven't been as popular as the complete engine minus carburetor. From the manufacturer's point of view, installing the heads and an intake manifold seals/covers the top of the engine and this is desirable from a shipping and storage standpoint.

The majority of the OEM/factory crate motors are shipped without a carburetor, which allows you to select one that meets your needs. Almost all crate motor intake manifolds are based on the single four-barrel design. In almost all cases, power curves and horsepower ratings are based on single four-barrel carburetion even if the engines are shipped with no intake manifold.

Since the mid-1990s, almost all production engines have been based on MPI—multi-point fuel injection. Many of these new production engines serve as the basis for today's crate motors and the factories offer manifolds for carbureted versions. For example, the Chrysler/Mopar Gen III Hemi engines, 5.7L and 6.1L in production, are offered with a single four-barrel option, plus an eight-barrel inline manifold is also available. Chevy/GM has also done this with the LS engine family. Typical crate motor offerings seem to be constantly changing, so check with the specific manufacturer for the latest details and up-to-date offerings. Check with GM Performance Parts, Ford Racing and Mopar Performance Parts for the latest information.

CRATE MOTOR SPECS

Mopar Crate Engines

Engine	Displ.	HP	Torque	C.R.	Heads	Cam	Induction
Hemi-Gen II	426	465	486	9.0	cast	Hyd	4bbl DP
	472	525	540	9.0	cast	Hyd	4bbl DP
	528	610	650	10.25	Alum.	Hyd-big	8bbl C-R
5.7 Hemi Gen III	5.7L	360	360	n/a	Alum.	Hyd-R	4bbl DP
440-style wedge	500	505	590	9.0	cast	Hyd-big	4bbl SP
Mopar Small-block	360	390	420	9.0	cast	Hyd.	4bbl SP
	406	435	470	9.0	cast	Hyd.	4bbl SP
	440	540	550	10.2	Alum.	Hyd.	4bbl SP

Edelbrock Crate Engines

Engine	Displ.	HP	Torque	C.R.	Heads	Cam	Induction
Chevy Small-block	350	320	382	9.0	Alum.	Hyd.	4bbl SP
	350	363	405	9.0	Alum.	Hyd-R	4bbl
	350	410	408	9.5	Alum.	Hyd-big	4bbl
	350	440	425	9.5	Alum.	Hyd-R	4bbl
	350	507	487	9.5	Alum.	Hyd-R	Supercharged
Ford Small-block	347	449	417	9.7	Alum	n/a	Dual 4bbl

Ford Crate Engines

Engine	Displ.	HP	Torque	C.R.	Heads	Cam	Induction
Ford Small-block	302	340	310	9.0	Alum.	Hyd-R	No intake
	302	390	360	10.0	Alum.	Hyd-R	No intake
	347	450	400	9.7	Alum.	Hyd-R	4bbl SP
	351	250	350	8.5	C.I.	Hyd	4bbl DP
	351	385	377	9.0	Alum.	Hyd-R	4bbl SP
	392	430	450	9.7	Alum	Hyd-R	4bbl SP
	392	475	495	10.0	Alum.	Hyd-R	4bbl SP

Chevy Crate Engines

Engine	Displ.	HP	Torque	C.R.	Heads	Cam	Induction
Chevy Small-block	350	290	326	8.5	C.I.	Hyd	No intake
	350	355	405	10.0	Alum.	Hyd-R	4bbl DP
	383	425	449	9.6	Alum.	Hyd-R	No intake
Chevy Big-block	427	430	444	10.0	Alum	Hyd-R	No intake
	454	440	500	9.6	Alum.	Hyd-R	4bbl DP
	502	450	550	8.75	C.I.	Hyd-R	4bbl DP
	572	620	650	9.6	Alum.	Hyd-R	4bbl SP

Note: Crate motors such as these listed above are constantly changing. As the hardware changes, the performance numbers evolve, so check with the manufacturer for the latest specs.